SCHOLASTIC

Algebra Readiness Made Easy

An ESSENTIAL Part of Every MATH Curriculum

Grade 5

CAROLE GREENES, CAROL FINDELL & MARY CAVANAGH

NEW YORK • TORONTO • LONDON • AUCKLAND • SYDNEY
MEXICO CITY • NEW DELHI • HONG KONG • BUENOS AIRES

Teaching Resources

Editor: Mela Ottaiano
Cover design by Jason Robinson
Interior design by Melinda Belter
Illustrations by Teresa Anderko

ISBN-13: 978-0-439-83936-5
ISBN-10: 0-439-83936-X

Printed in China.

1 2 3 4 5 6 7 8 9 10   15 14 13 12 11 10 09 08

Table of Contents

Introduction

Welcome to *Algebra Readiness Made Easy*! This book is designed to help you introduce students to problem-solving strategies and algebraic-reasoning techniques, to give them practice with major number concepts and skills, and to motivate them to write and talk about big ideas in mathematics. It also sets the stage for the formal study of algebra in the upper grades.

Algebra Standards

The National Council of Teachers of Mathematics identifies algebra as one of the five major content areas of the mathematics curriculum to be studied by students in *all* grades (NCTM, 2000). The council emphasizes that early and regular experience with the key ideas of algebra helps students make the transition into the more formal study of algebra in late middle school or high school. This view is consistent with the general theory of learning—that understanding is enhanced when connections are made between what is new and what was previously studied. The key algebraic concepts developed in this book are:

- representing quantitative relationships with symbols
- writing and solving equations
- solving equations with one or more variables
- replacing unknowns with their values
- solving for the values of unknowns
- solving two or three equations with two or three unknowns
- exploring equality
- exploring variables that represent varying quantities
- describing the functional relationship between two numbers

Building Key Math Skills

NCTM also identifies problem solving as a key process skill, and the teaching of strategies and methods of reasoning to solve problems as a major part of the mathematics curriculum for students of all ages. The problem-solving model first described in 1957 by the renowned mathematician George Polya has been adopted by teachers and instructional developers nationwide and provides the framework for the problem-solving focus of this book. All the problems contained here require students to interpret data displays—such as text, charts, diagrams, pictures, and tables—and answer questions about them. As they work on the problems, students learn and practice the following problem-solving strategies:

- making lists of possible solutions, and testing those solutions
- identifying, describing, and generalizing patterns
- working backward
- reasoning logically
- reasoning proportionally

The development of problem-solving strategies and algebraic concepts is linked to the development of number concepts and skills. As students solve the problems in this book, they'll practice computing, applying concepts of place value and number theory, reasoning about the magnitudes of numbers, and more.

Throughout this book, we emphasize the language of mathematics. This language includes terminology (e.g., *odd number, variable*) as well as symbols (e.g., $\geq$, $\leq$). Students will see the language in the problems and illustrations and use the language in their discussions and written descriptions of their solution processes.

How to Use This Book

Inside this book you'll find six problem sets—each composed of nine problems featuring the same type of data display (e.g., diagrams, scales, and arrays of numbers)—that focus on one or more problem-solving strategies and algebraic concepts. Each set opens with an overview of the type of problems/tasks in the set, the algebra and problem-solving focus, the number concepts or skills needed to solve the problems, the math language emphasized in the problems, and guiding questions to be used with the first two problems of the set to help students grasp the key concepts and strategies.

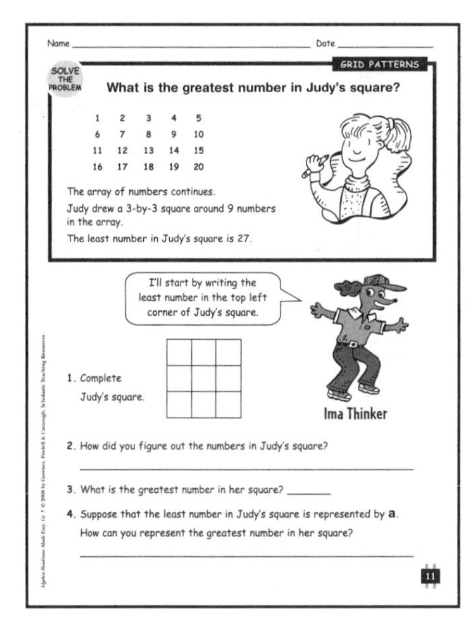

Name ______________ Date ______

SOLVE THE PROBLEM

GRID PATTERNS

What is the greatest number in Judy's square?

1	2	3	4	5
6	7	8	9	10
11	12	13	14	15
16	17	18	19	20

The array of numbers continues.
Judy drew a 3-by-3 square around 9 numbers in the array.
The least number in Judy's square is 27.

I'll start by writing the least number in the top left corner of Judy's square.

Ima Thinker

1. Complete Judy's square.

2. How did you figure out the numbers in Judy's square?

3. What is the greatest number in her square? ______

4. Suppose that the least number in Judy's square is represented by **a**. How can you represent the greatest number in her square?

11

Algebra Readiness Made Easy: Gr. 5 © 2008 by Greenes, Findell & Cavanagh, Scholastic Teaching Resources

The first two problems in each set are designed to be discussed and solved in a whole-class setting. The first, "Solve the Problem," introduces students to the type of display and problem they will encounter in the rest of the set. We suggest that you have students work on this first problem individually or in pairs before you engage in any formal instruction. Encourage students to wrestle with the problem and come up with some strategies they might use to solve it. Then gather students together and use the guiding questions provided to help them discover key mathematical relationships and understand the special vocabulary used in the

problem. This whole-class discussion will enhance student understanding and success with the problem-solving strategies and algebraic concepts in each problem set.

The second problem, "Make the Case," comes as an overhead transparency and uses a multiple-choice format. Three different characters offer possible solutions to the problem. Students have to determine which character—Sally Soccer, Buddy Basketball, Bobby Baseball—has the correct answer. Before they can identify the correct solution, students have to solve the problem themselves and analyze each of the responses. Invite them to speculate about why the other two characters got the wrong answers. (Note: Although we offer a rationale for each wrong answer, other explanations are possible.) As students justify their choices in the "Make the Case" problems, they gain greater experience using math language.

While working on these first two problems, it is important to encourage students to talk about their observations and hypotheses. This talk provides a window into what students do and do not understand. Working on "Solve the Problem" and "Make the Case" should take approximately one math period.

The rest of the problems in each set are sequenced by difficulty. All problems feature a series of questions that involve analyses of the data display. In the first three or four problems of each set, problem-solving "guru" Ima Thinker provides hints about how to begin solving the problems. No hints are provided for the rest of the problems. If students have difficulty solving these latter problems, you might want to write "Ima" hints for each of them or ask students to develop hints before beginning to solve the problems. An answer key is provided at the back of the book.

The problem sets are independent of one another and may be used in any order and incorporated into the regular mathematics curriculum at whatever point makes sense. We recommend that you work with each problem set in its entirety before moving on to the next one. Once you and your students work through the first two problems, you can assign problems 1 through 7 for students to do on their own or in pairs. You may wish to have them complete the problems during class or for homework.

Using the Transparencies

In addition to the reproducible problem sets, you'll find ten overhead transparencies at the back of this book. (Black-line masters of all transparencies also appear in the book.) The first six transparencies are reproductions of the "Make the Case" problems, to help you in leading a whole-class discussion of each problem.

The remaining four transparencies are designed to be used together. Three of these transparencies feature six problems, one from each of the problem sets. Cut these three transparencies in half and overlay each problem on the Problem-Solving Transparency. Then invite students to apply our three-step problem-solving process:

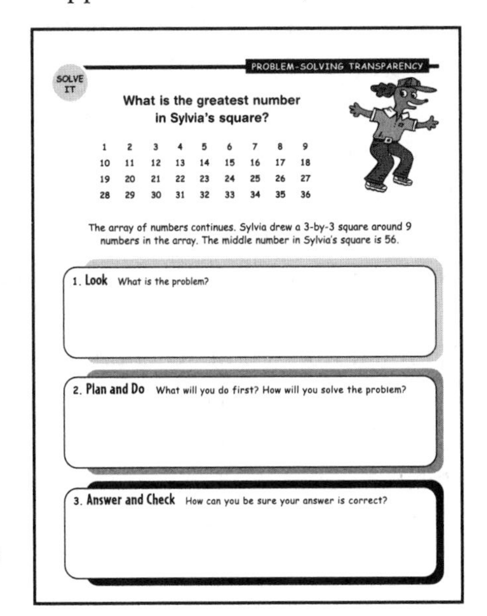
PROBLEM-SOLVING TRANSPARENCY

SOLVE IT

What is the greatest number in Sylvia's square?

1	2	3	4	5	6	7	8	9
10	11	12	13	14	15	16	17	18
19	20	21	22	23	24	25	26	27
28	29	30	31	32	33	34	35	36

The array of numbers continues. Sylvia drew a 3-by-3 square around 9 numbers in the array. The middle number in Sylvia's square is 56.

1. **Look** What is the problem?

2. **Plan and Do** What will you do first? How will you solve the problem?

3. **Answer and Check** How can you be sure your answer is correct?

1) **Look:** What is the problem? What information do you have? What information do you need?

2) **Plan and Do:** How will you solve the problem? What strategies will you use? What will you do first? What's the next step? What comes after that?

3) **Answer and Check:** What is the answer? How can you be sure that your answer is correct?

These problem-solving transparencies encourage writing about mathematics and may be used at any time. They are particularly effective when used as culminating activities for the set of problems.

References

Greenes, Carole, & Carol Findell. (Eds.). (2005). *Developing students' algebraic reasoning abilities.* (Vol. 3 in the NCSM Monograph Series.) Boston, MA: Houghton Mifflin.

Greenes, Carole, & Carol Findell. (2005). *Groundworks: Algebraic thinking.* Chicago: Wright Group/McGraw Hill.

Greenes, Carole, & Carol Findell. (2007, 2008). *Problem solving think tanks.* Brisbane, Australia: Origo Education.

Moses, Barbara. (Ed.). (1999). *Algebraic thinking, grades K–12: Readings from NCTM's school-based journals and other publications.* Reston, VA: National Council of Teachers of Mathematics.

National Council of Teachers of Mathematics. (2000). *Principles and standards for school mathematics.* Reston, VA: National Council of Teachers of Mathematics.

National Council of Teachers of Mathematics. (2008). *Algebra and algebraic thinking in school mathematics,* 2008 Yearbook. (C. Greenes, Ed.) Reston, VA: National Council of Teachers of Mathematics.

Polya, George. (1957). *How to solve it.* Princeton, NJ: Princeton University Press.

Grid Patterns

Overview

Students identify relationships among numbers in a rectangular array of counting numbers.

Algebra Focus

Explore variables as representing varying quantities • Describe the functional relationship between the numbers in a 3-by-3 section of an array of counting numbers

Problem-Solving Strategies

Describe parts of patterns • Generalize pattern relationships

Related Math Skills ≤ ≥ X ÷

Compute with counting numbers

Math Language

Array • Grid • Greatest number • Least number •
Middle number (in a 3-by-3 square) • Represent • 3-by-3 square

Introducing the Problem Set

Make photocopies of "Solve the Problem: Grid Patterns" (page 11) and distribute to students. Have students work in pairs, encouraging them to discuss strategies they might use to solve the problem. You may want to walk around and listen in on some of their discussions. After a few minutes, display the problem on the board (or on the overhead if you made a transparency) and use the following questions to guide a whole-class
discussion on how to solve the problem:

- What is the last number in each row? *(a multiple of 5)*
- What number is just to the right of 12? *(13)*

- What number is just below 13? *(18)* And just below 18? *(23)*
- In this array, how can you figure out the number that is just below a given number? *(Add 5 to the given number.)*
- How can you figure out the greatest number in Judy's square? *(The greatest number is 2 numbers to the right and 2 numbers below the least number. The greatest number is 27 + 1 + 1 + 5 + 5 or 39.)*

Work together as a class to answer the questions in "Solve the Problem: Grid Patterns."

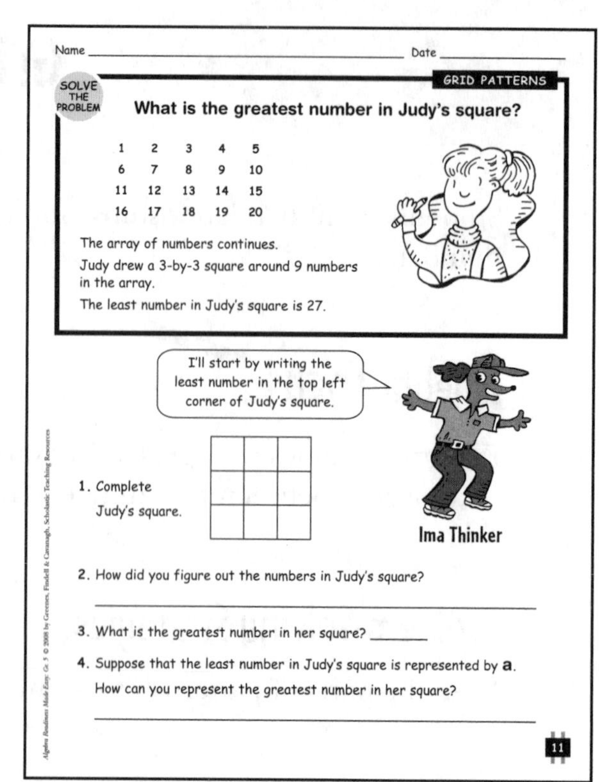

Name ______________________ Date ________

SOLVE THE PROBLEM — GRID PATTERNS

What is the greatest number in Judy's square?

1	2	3	4	5
6	7	8	9	10
11	12	13	14	15
16	17	18	19	20

The array of numbers continues.
Judy drew a 3-by-3 square around 9 numbers in the array.
The least number in Judy's square is 27.

1. Complete Judy's square.
2. How did you figure out the numbers in Judy's square? ______________________
3. What is the greatest number in her square? ________
4. Suppose that the least number in Judy's square is represented by **a**. How can you represent the greatest number in her square? ______________________

Math Chat With the Transparency

Display the "Make the Case: Grid Patterns" transparency on the overhead. Before students can decide which character is "on the ball," they need to figure out the answer to the problem. Encourage students to work in pairs to solve the problem, then bring the class together for another whole-class discussion. Ask:

- Who has the right answer? *(Buddy)*
- How did you figure it out? *(In this array, each number in a column is 4 less than the number below it. The least number in the square is two numbers to the left and two numbers above the greatest number. The least number is 39 – 1 – 1 – 4 – 4, or 29.)*
- How do you think Sally got the answer 31? *(She probably counted back by ones to get to the least number; 39 – 8 = 31.)*
- How do you think Bobby got the answer of 49? *(He probably thought about 39 as the least number and figured out the greatest number. He started at 39 and added 1 + 1 + 4 + 4.)*

Name ______________________ Date ________

MAKE THE CASE — GRID PATTERNS

What is the least number in Kevin's square?

1	2	3	4
5	6	7	8
9	10	11	12
13	14	15	16
17	18	19	20

The array of numbers continues.
Kevin drew a 3-by-3 square around 9 numbers in the array.
The greatest number in Kevin's square is 39.

Who is on the ball?

Name ______________________________ Date ______________

SOLVE THE PROBLEM

What is the greatest number in Judy's square?

1	2	3	4	5
6	7	8	9	10
11	12	13	14	15
16	17	18	19	20

The array of numbers continues.

Judy drew a 3-by-3 square around 9 numbers in the array.

The least number in Judy's square is 27.

I'll start by writing the least number in the top left corner of Judy's square.

Ima Thinker

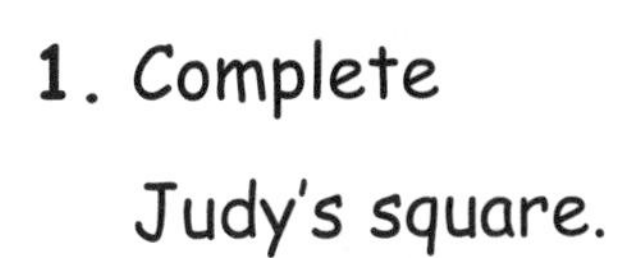

1. Complete Judy's square.

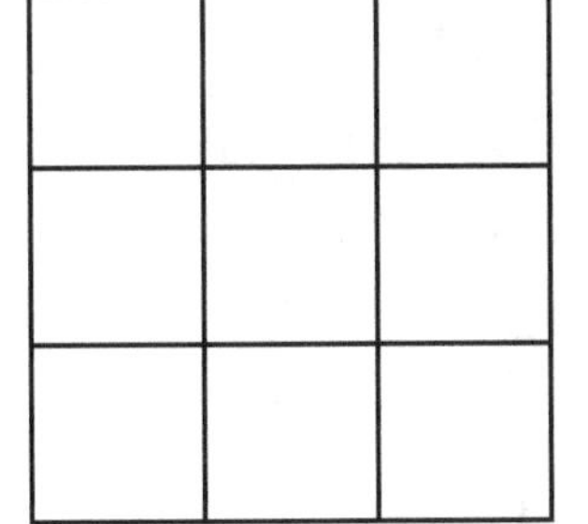

2. How did you figure out the numbers in Judy's square?

____

3. What is the greatest number in her square? ________

4. Suppose that the least number in Judy's square is represented by **a**.

   How can you represent the greatest number in her square?

____

Name ______________________________ Date ______________

MAKE THE CASE

What is the least number in Kevin's square?

1	2	3	4
5	6	7	8
9	10	11	12
13	14	15	16
17	18	19	20

The array of numbers continues.

Kevin drew a 3-by-3 square around 9 numbers in the array.

The greatest number in Kevin's square is 39.

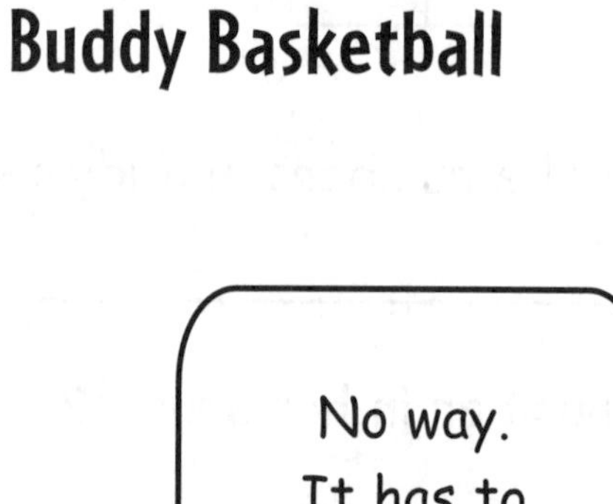

Who is on the ball?

Name ______________________________________ Date ________________

GRID PATTERNS

PROBLEM 1

What is the least number in Lara's square?

1	2	3
4	5	6
7	8	9
10	11	12
13	14	15
16	17	18

The array of numbers continues.

Lara drew a 3-by-3 square around 9 numbers in the array.

The greatest number in Lara's square is 27.

I'll start by writing the greatest number in the bottom right corner of Lara's square.

Ima Thinker

1. Complete Lara's square.

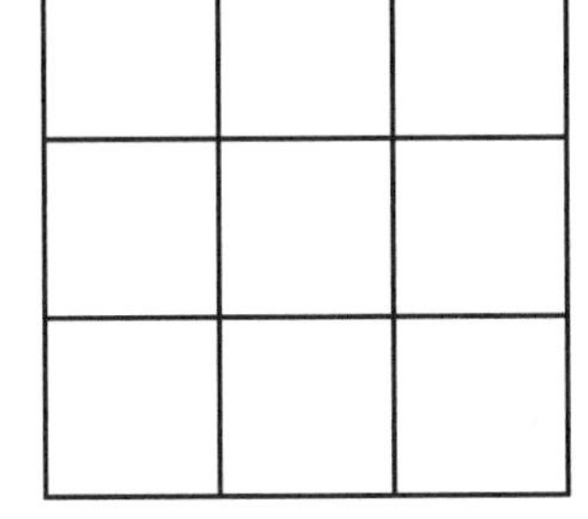

2. How did you figure out the numbers in Lara's square?

____

3. What is the least number in her square? ________

4. Suppose that the greatest number in Lara's square is represented by **b**.

   How can you represent the least number in her square?

____

Name ______________________________ Date ______________

PROBLEM 2

GRID PATTERNS

What is the least number in Morey's square?

1	2	3	4	5	6	7	8	9	10
11	12	13	14	15	16	17	18	19	20
21	22	23	24	25	26	27	28	29	30
31	32	33	34	35	36	37	38	39	40

The array of numbers continues.

Morey drew a 3-by-3 square around 9 numbers in the array.

The greatest number in Morey's square is 48.

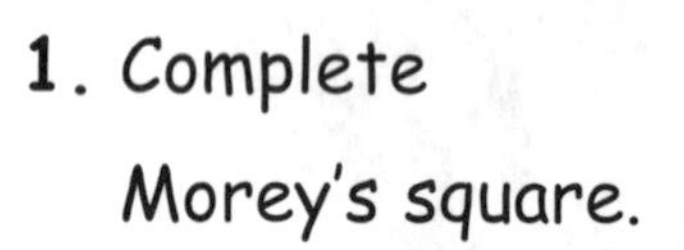

1. Complete Morey's square.

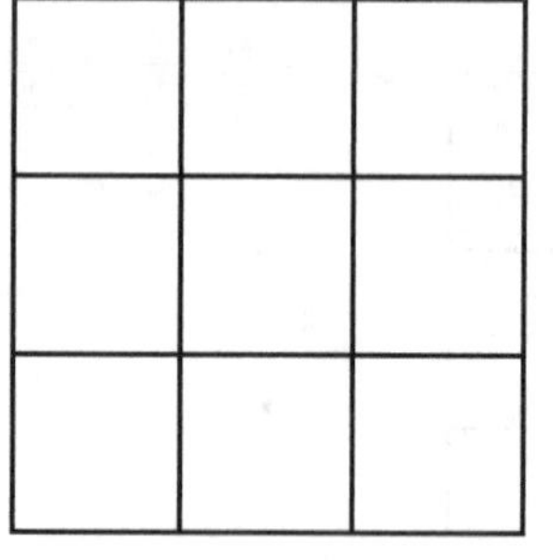

2. How did you figure out the numbers in Morey's square?

____

3. What is the least number in his square? ________

4. Suppose that the greatest number in Morey's square is represented by **C**. How can you represent the least number in his square?

____

Name ________________________ Date ____________

PROBLEM 3

What is the greatest number in Nadia's square?

1	2	3	4	5	6
7	8	9	10	11	12
13	14	15	16	17	18
19	20	21	22	23	24

The array of numbers continues.

Nadia drew a 3-by-3 square around 9 numbers in the array.

The least number in Nadia's square is 26.

I'll start by writing the least number in the top left corner of Nadia's square.

Ima Thinker

1. Complete Nadia's square.

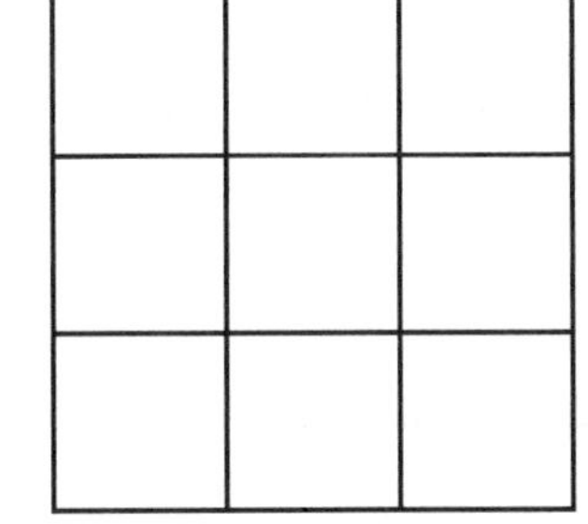

2. How did you figure out the numbers in Nadia's square?

____

3. What is the greatest number in her square? ________

4. Suppose that the least number in Lara's square is represented by **d**. How can you represent the greatest number in her square?

____

Name ____________________ Date __________

PROBLEM 4

What is the greatest number in Oliver's square?

1	2	3	4	5	6	7	8
9	10	11	12	13	14	15	16
17	18	19	20	21	22	23	24
25	26	27	28	29	30	31	32

The array of numbers continues.

Oliver drew a 3-by-3 square around 9 numbers in the array.

The least number in Oliver's square is 34.

1. Complete Oliver's square.

2. How did you figure out the numbers in Oliver's square?

____

3. What is the greatest number in his square? ________

4. Suppose that the least number in Oliver's square is represented by **e**.

   How can you represent the greatest number in his square?

____

Name ______________________ Date ______________

GRID PATTERNS

PROBLEM 5

What is the least number in Paula's square?

1	2	3	4	5
6	7	8	9	10
11	12	13	14	15
16	17	18	19	20

The array of numbers continues.

Paula drew a 3-by-3 square around 9 numbers in the array.

The greatest number in Paula's square is 58.

1. Complete Paula's square.

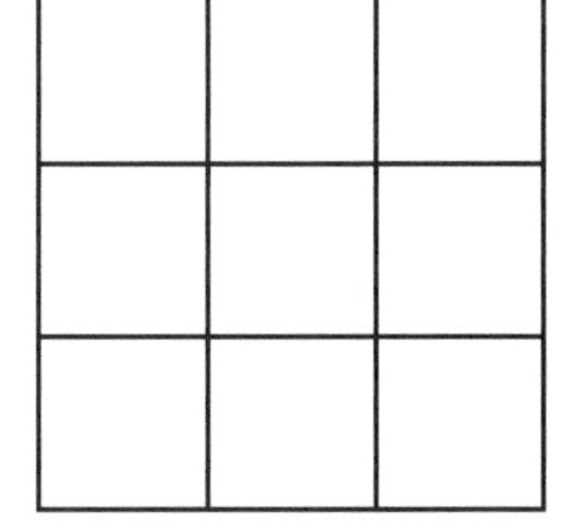

2. What is the least number in her square?

____

3. Suppose that the greatest number in Paula's square is represented by **f**. How can you represent the least number in her square?

____

4. Suppose that Paula draws a different 3-by-3 square and the greatest number is 47. What is the least number? _______

Name ______________________ Date ____________

PROBLEM 6

GRID PATTERNS

What is the greatest number in Quent's square?

1	2	3	4
5	6	7	8
9	10	11	12
13	14	15	16
17	18	19	20

The array of numbers continues.

Quent drew a 3-by-3 square around 9 numbers in the array.

The middle number in Quent's square is 51.

1. Complete Quent's square.

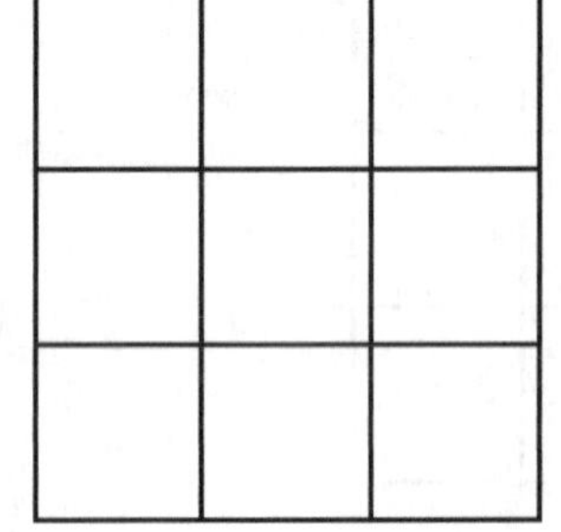

2. What is the greatest number in his square?

____

3. Suppose that the middle number in Quent's square is represented by **g**. How can you represent the greatest number in his square?

____

4. Suppose that Quent draws a different 3-by-3 square and the middle number is 59. What is the greatest number? ________

Name ______________________________ Date ______________

PROBLEM 7

GRID PATTERNS

What is the least number in Richard's square?

1	2	3	4	5	6
7	8	9	10	11	12
13	14	15	16	17	18
19	20	21	22	23	24

The array of numbers continues.

Richard drew a 3-by-3 square around 9 numbers in the array.

The middle number in Richard's square is 45.

1. Complete Richard's square.

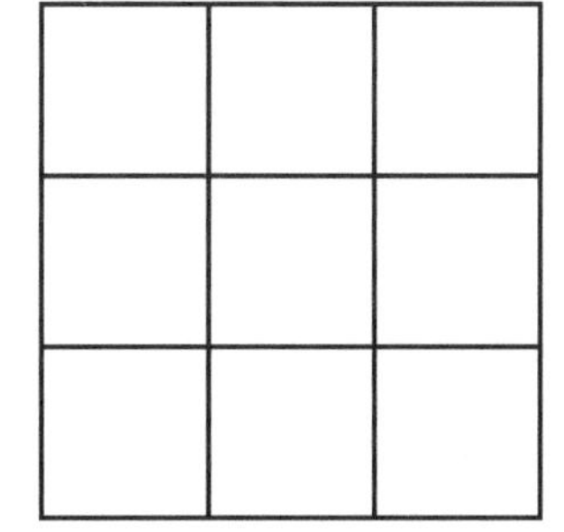

2. What is the least number in his square?

____

3. Suppose that the middle number in Richard's square is represented by **h**. How can you represent the least number in his square?

____

4. Suppose that Richard draws a different 3-by-3 square and the middle number is 62. What is the least number? ________

Dollar Dilemma

Overview

Students interpret mathematical relationships and work backward through sets of clues to determine costs of various animal accessories.

Algebra Focus

Interpret quantitative relationships • Write and solve equations

Problem-Solving Strategies

Eliminate candidates from a list • Use logical reasoning

Related Math Skills $\leq$ $\geq$ $\times$ $\div$

Compute fractional parts of whole numbers of dollars ($\frac{1}{2}$, $\frac{1}{3}$, $\frac{1}{4}$, $\frac{1}{5}$, $\frac{1}{6}$, $\frac{1}{10}$)

Math Language

More than • Less than • $\frac{1}{2}$, $\frac{1}{3}$, $\frac{1}{4}$, $\frac{1}{5}$, $\frac{1}{6}$, $\frac{1}{10}$ • Twice as much • Half as much

Introducing the Problem Set

Make photocopies of "Solve the Problem: Dollar Dilemma" (page 22) and distribute to students. Have students work in pairs, encouraging them to discuss strategies they might use to solve the problem. You may want to walk around and listen in on some of their discussions. After a few minutes, display the problem on the board (or on the overhead if you made a transparency) and use the following questions to guide a whole-class discussion on how to solve the problem:

- Which fact did Ima use first? *(Fact D)* Why did Ima start with that fact? *(Fact D is the only fact that gives the cost of one cat's earphones. All other facts need more information than is given. Knowing that Smudge's earphones cost $10, the cost of Tigger's earphones can be computed, and so on.)*
- What is the cost of Tigger's earphones? *($2)* How did you figure out the cost of Tigger's earphones? *(* $\frac{1}{5} \times 10 = \$2$ *)*

- If you know the cost of Tigger's earphones, how can you figure out the cost of Cookie's earphones? *(8 x 2 = $16)*
- How can you figure out the cost of Mouser's earphones? *(1 + [½ x 16] = $9)*

Work together as a class to answer the questions in "Solve the Problem: Dollar Dilemma."

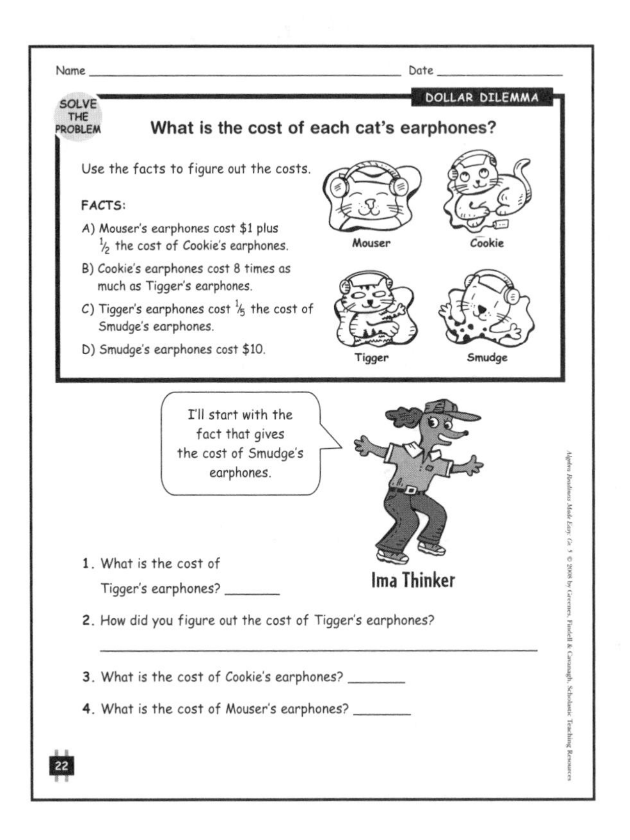
Name ______ Date ______

SOLVE THE PROBLEM — DOLLAR DILEMMA

What is the cost of each cat's earphones?

Use the facts to figure out the costs.

FACTS:

A) Mouser's earphones cost $1 plus ½ the cost of Cookie's earphones.

B) Cookie's earphones cost 8 times as much as Tigger's earphones.

C) Tigger's earphones cost ⅕ the cost of Smudge's earphones.

D) Smudge's earphones cost $10.

1. What is the cost of Tigger's earphones? ______
2. How did you figure out the cost of Tigger's earphones? ______
3. What is the cost of Cookie's earphones? ______
4. What is the cost of Mouser's earphones? ______

22

Math Chat With the Transparency

Display the "Make the Case: Dollar Dilemma" transparency on the overhead. Before students can decide which character is "on the ball," they need to figure out the answer to the problem. Encourage students to work in pairs to solve the problem, then bring the class together for another whole-class discussion. Ask:

- Who has the right answer? *(Bobby)*
- How did you figure it out? *(Work backward. From Fact D, Bubble's jump rope is $7. From Fact C, Larry's jump rope is 7 – 1, or $6. From Fact B, Fin's jump rope is 2 + [½ x 6], or $5. From Fact A, Bob's jump rope is 3 + [⅕ x 5], or $4.)*
- How do you think Sally got her answer of $8? *(She probably used Facts D, C, and B correctly to figure out the cost of Fin's jump rope which is $5. She then added $3 to that amount to get the cost of Bob's jump rope. She forgot to take ⅕ of the cost of Fin's jump rope before adding the $3.)*
- How do you think Buddy got the answer of $5? *(He probably added the dollar amounts showing in Facts B, C, and D ($2, $1, and $7) and then applied Fact A: 3 + [⅕ x 10], or $5.)*

Name ______ Date ______

MAKE THE CASE — DOLLAR DILEMMA

What is the cost of Bob's jump rope?

Use the facts to figure out the costs.

FACTS:

A) Bob's jump rope costs $3 plus ⅕ the cost of Fin's rope.

B) Fin's jump rope costs $2 plus ½ the cost of Larry's jump rope.

C) Larry's jump rope costs $1 less than Bubbles' jump rope.

D) Bubbles' jump rope costs $7.

Who is on the ball?

23

Name ______________________________ Date ______________

SOLVE THE PROBLEM

DOLLAR DILEMMA

What is the cost of each cat's earphones?

Use the facts to figure out the costs.

Mouser

Cookie

Tigger

Smudge

FACTS:

A) Mouser's earphones cost $1 plus $\frac{1}{2}$ the cost of Cookie's earphones.

B) Cookie's earphones cost 8 times as much as Tigger's earphones.

C) Tigger's earphones cost $\frac{1}{5}$ the cost of Smudge's earphones.

D) Smudge's earphones cost $10.

I'll start with the fact that gives the cost of Smudge's earphones.

Ima Thinker

1. What is the cost of Tigger's earphones? ________

2. How did you figure out the cost of Tigger's earphones?

____

3. What is the cost of Cookie's earphones? ________

4. What is the cost of Mouser's earphones? ________

Name ______________________________ Date ______________

DOLLAR DILEMMA

MAKE THE CASE

What is the cost of Bob's jump rope?

Use the facts to figure out the costs.

FACTS:

A) Bob's jump rope costs \$3 plus $\frac{1}{5}$ the cost of Fin's rope.

B) Fin's jump rope costs \$2 plus $\frac{1}{2}$ the cost of Larry's jump rope.

C) Larry's jump rope costs \$1 less than Bubbles' jump rope.

D) Bubbles' jump rope costs \$7.

Who is on the ball?

Name ____________________ Date __________

PROBLEM 1

DOLLAR DILEMMA

What is the cost of each dog's badminton racquet?

Use the facts to figure out the costs.

FACTS:

A) Skipper's racquet costs $2 less than Pooch's racquet.

B) Pooch's racquet costs $3 plus $\frac{1}{2}$ the cost of Callie's racquet.

C) Callie's racquet costs $2 more than Holly's racquet.

D) Holly's racquet costs $6.

1. What is the cost of Callie's racquet? ________

2. How did you figure out the cost of Callie's racquet?

____

3. What is the cost of Pooch's racquet? ________

4. What is the cost of Skipper's racquet? ________

Name ______________________ Date ______________

PROBLEM 2

DOLLAR DILEMMA

What is the cost of each dog's flip-flops?

Use the facts to figure out the costs.

Curly

Ziggy

Squealy

Porky

FACTS:

A) Curly's flip-flops cost $7 plus twice the cost of Ziggy's flip-flops.

B) Ziggy's flip-flops cost half as much as Squealy's flip-flops.

C) Squealy's flip-flops cost $8 plus $\frac{1}{4}$ the cost of Porky's flip-flops.

D) Porky's flip-flops cost $16.

I'll start with that fact that gives the cost of Porky's flip-flops.

Ima Thinker

1. What is the cost of Squealy's flip-flops? ________

2. How did you figure out the cost of Squealy's flip-flops?

____

3. What is the cost of Ziggy's flip-flops? ________

4. What is the cost of Curly's flip-flops? ________

Name ______________________________ Date ______________

PROBLEM 3

DOLLAR DILEMMA

What is the cost of each goose's skateboard?

Use the facts to figure out the costs.

FACTS:

A) Dobbin's skateboard costs $63 minus the cost of Mandy's skateboard.

B) Mandy's skateboard costs $6 plus $\frac{1}{2}$ the cost of Dandy's skateboard.

C) Dandy's skateboard costs $32 plus $\frac{1}{10}$ the cost of Sebastian's skateboard.

D) Sebastian's skateboard costs $40.

Dobbin

Mandy

Dandy

Sebastian

I'll start with that fact that gives the cost of Sebastian's skateboard.

Ima Thinker

1. What is the cost of Dandy's skateboard? ________

2. How did you figure out the cost of Dandy's skateboard?

3. What is the cost of Mandy's skateboard? ________

4. What is the cost of Dobbin's skateboard? ________

Name ________________________________ Date ______________

PROBLEM 4

DOLLAR DILEMMA

What is the cost of each giraffe's baseball glove?

Use the facts to figure out the costs.

FACTS:

A) Gerry's glove costs \$10 plus $\frac{1}{2}$ the cost of Celia's glove.

B) Celia's glove costs \$2 more than Shorty's glove.

C) Shorty's glove costs \$26 plus $\frac{1}{6}$ the cost of Jack's glove.

D) Jack's glove costs \$24.

Gerry

Celia

Shorty

Jack

1. What is the cost of Shorty's glove? ________

2. How did you figure out the cost of Shorty's glove?

____

3. What is the cost of Celia's glove? ________

4. What is the cost of Gerry's glove? ________

Name ______________________ Date ____________

PROBLEM 5

DOLLAR DILEMMA

What is the cost of each cow's skis?

Use the facts to figure out the costs.

FACTS:

A) Bertha's skis cost $30 more than Brownie's skis.

B) Brownie's skis cost $100 plus $\frac{1}{10}$ the cost of Elsie's skis.

C) Elsie's skis cost $60 plus $\frac{1}{3}$ the cost of Splotch's skis.

D) Splotch's skis cost $120.

Bertha

Brownie

Elsie

Splotch

1. What is the cost of Elsie's skis? ________

2. How did you figure out the cost of Elsie's skis?

____

3. What is the cost of Brownie's skis? ________

4. What is the cost of Bertha's skis? ________

Name ______________________ Date ____________

DOLLAR DILEMMA

PROBLEM 6

What is the cost of each sheep's bicycle helmet?

Use the facts to figure out the costs.

FACTS:

A) Wooly's helmet costs $17 plus $\frac{1}{2}$ the cost of Lammy's helmet.

B) Lammy's helmet costs $4 less than Sweetie's helmet.

C) Sweetie's helmet costs $\frac{1}{2}$ the total cost of Allie's and Blackie's helmets.

D) Allie's helmet costs $8 plus $\frac{1}{3}$ the cost of Blackie's helmet.

E) Blackie's helmet costs $33.

Blackie

Allie

Wooly

Sweetie

Lammy

1. What is the cost of Allie's helmet? ________

2. What is the cost of Sweetie's helmet? ________

3. What is the cost of Lammy's helmet? ________

4. What is the cost of Wooly's helmet? ________

Name ______________________ Date __________

PROBLEM 7

DOLLAR DILEMMA

What is the cost of each mouse's sunglasses?

Use the facts to figure out the costs.

FACTS:

A) Albert's glasses cost $\frac{1}{2}$ the total cost of Betsy's and Eric's glasses.

B) Betsy's glasses cost \$12 plus $\frac{1}{3}$ the cost of Eric's glasses.

C) Eric's glasses cost \$13 plus $\frac{1}{5}$ the cost of Darin's glasses.

D) Darin's glasses cost \$3 plus $\frac{1}{2}$ the cost of Cookie's glasses.

E) Cookie's glasses cost \$14.

1. What is the cost of Darin's glasses? ________
2. What is the cost of Eric's glasses? ________
3. What is the cost of Betsy's glasses? ________
4. What is the cost of Albert's glasses? ________

Birthday Boggle

Overview

Students use clues and reason logically to figure out birth dates of famous people. The birth dates, the unknowns, are represented by letters.

Algebra Focus

Solve for values of unknowns • Replace letters with their values

Problem-Solving Strategies

Make a list of possible solutions • Test possible solutions with clues • Use logical reasoning

Related Math Skills ≤ ≥ X ÷

Compute with whole numbers • Identify factors of numbers • Identify odd and even numbers

Math Language

Difference • Digit • Even number • Factor • Odd number • Product • Remainder • Sum • Symbols: Not equal to ≠, Less than <, Greater than >, Less than or equal to ≤, Greater than or equal to ≥

Introducing the Problem Set

Make photocopies of "Solve the Problem: Birthday Boggle" (page 33) and distribute to students. Have students work in pairs, encouraging them to discuss strategies they might use to solve the problem. You may want to walk around and listen in on some of their discussions. After a few minutes, display the problem on the board (or on the overhead if you made a transparency) and use the following questions to guide a whole-class discussion on how to solve the problem:

- Look at Clue 1. What is the least number that M can be? *(5)* How did you figure it out? *(If $M + M + M$ is greater than or equal to 15, then M is greater than or equal to $15 \div 3$, or 5.)*

- Look at Clue 2. What is the greatest number that M can be? *(11)* How did you figure it out? *(M is less than 12 and the greatest whole number less than 12 is 11.)*
- What numbers are on Ima's list? *(5, 6, 7, 8, 9, 10, and 11.)*
- What numbers on Ima's list are eliminated by Clue 3? *(All numbers except for 5 and 10.)*
- Which numbers does Clue 4 eliminate? *(10)*
- How can you check your answer? *(Replace each M in the clues with its value. Be sure that the statements are true.)*

Work together as a class to answer the questions in "Solve the Problem: Birthday Boggle."

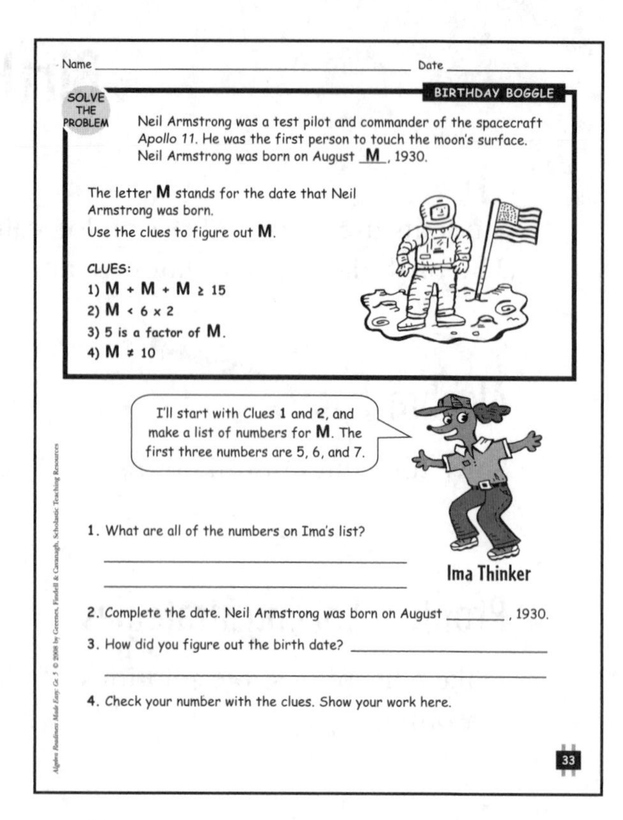

Name ______________________ Date ____________

SOLVE THE PROBLEM — BIRTHDAY BOGGLE

Neil Armstrong was a test pilot and commander of the spacecraft *Apollo 11*. He was the first person to touch the moon's surface. Neil Armstrong was born on August __M__, 1930.

The letter **M** stands for the date that Neil Armstrong was born.
Use the clues to figure out **M**.

CLUES:
1) $M + M + M \geq 15$
2) $M < 6 \times 2$
3) 5 is a factor of **M**.
4) $M \neq 10$

1. What are all of the numbers on Ima's list? ______________________
2. Complete the date. Neil Armstrong was born on August ________, 1930.
3. How did you figure out the birth date? ______________________
4. Check your number with the clues. Show your work here.

33

Math Chat With the Transparency

Display the "Make the Case: Birthday Boggle" transparency on the overhead. Before students can decide which character is "on the ball," they need to figure out the answer to the problem. Encourage students to work in pairs to solve the problem, then bring the class together for another whole-class discussion. Ask:

- Who has the right answer? *(Bobby)*
- When was Mary Cassatt born? *(May 22, 1944)*
- How did you figure out the value of P? *(From Clue 1, P can be 1, 2, 3, . . ., or 25. Clue 2 eliminates all odd numbers. Clue 3 eliminates all even numbers except for 20, 22, and 24. Clue 4 eliminates 20 and 24. So, P = 22.)*
- How do you think Sally got 20? *(20 fits clues 1, 2, and 3. Sally probably forgot to use Clue 4.)*
- How do you think Buddy got 24? *(24 fits clues 1, 2, and 3. Buddy probably forgot to use Clue 4.)*

Name ______________________ Date ____________

MAKE THE CASE — BIRTHDAY BOGGLE

Mary Cassatt was a painter and printmaker. She is most famous for her paintings of children. Mary Cassatt was born on May __P__, 1844.

The letter **P** stands for the birth date.
Use the clues to figure out **P**.

CLUES:
1) $P \leq 5 \times 5$
2) $P \div 2$ has a zero remainder.
3) The sum of the tens and ones digits of **P** is an even number.
4) The difference between the tens and ones digits is less than 2.

Who is on the ball?

34

Name ______________________________ Date ______________

BIRTHDAY BOGGLE

SOLVE THE PROBLEM

Neil Armstrong was a test pilot and commander of the spacecraft *Apollo 11*. He was the first person to touch the moon's surface. Neil Armstrong was born on August __M__, 1930.

The letter **M** stands for the date that Neil Armstrong was born.

Use the clues to figure out **M**.

CLUES:

1) $M + M + M \geq 15$
2) $M < 6 \times 2$
3) 5 is a factor of **M**.
4) $M \neq 10$

I'll start with Clues **1** and **2**, and make a list of numbers for **M**. The first three numbers are 5, 6, and 7.

1. What are all of the numbers on Ima's list?

____

____

2. Complete the date. Neil Armstrong was born on August ________, 1930.

3. How did you figure out the birth date? ______________________________

____

4. Check your number with the clues. Show your work here.

Name ______________________ Date ____________

BIRTHDAY BOGGLE

MAKE THE CASE

Mary Cassatt was a painter and printmaker. She is most famous for her paintings of children. Mary Cassatt was born on May __P__, 1844.

The letter **P** stands for the birth date. Use the clues to figure out **P**.

CLUES:

1) **P** ≤ 5 × 5
2) **P** ÷ 2 has a zero remainder.
3) The sum of the tens and ones digits of **P** is an even number.
4) The difference between the tens and ones digits is less than 2.

Buddy Basketball

Sally Soccer

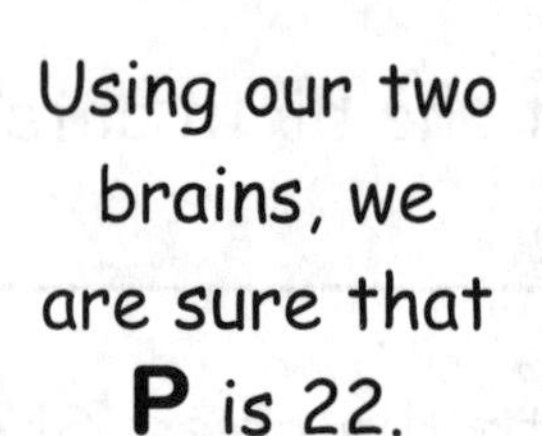

Bobby Baseball

Who is on the ball?

Name ________________________ Date ______________

PROBLEM 1 — **BIRTHDAY BOGGLE**

Betsy Ross was a seamstress who sewed the first American Flag in 1776. Betsy was born on January __**R**__, 1752.

The letter **R** stands for the date that Betsy Ross was born.

Use the clues to figure out **R**.

CLUES:

1) $R \times R \leq 25$
2) **R** is an odd number.
3) 5 is not a factor of **R**.
4) $R \neq 27 \div 9$

1. What are all of the numbers on Ima's list?

____

____

2. Complete the date. Betsy Ross was born on January ________ , 1752.

3. How did you figure out the birth date? ________________________________

____

4. Check your number with the clues. Show your work here.

Name ______________________________ Date ______________

PROBLEM 2 | **BIRTHDAY BOGGLE**

Samuel Morse was a professional artist and inventor of a code to send messages. The code uses dots and dashes to represent letters, numbers, question marks, commas, and periods. Samuel Morse was born on April __**C**__, 1791.

The letter **C** stands for the date that Samuel Morse was born.

Use the clues to figure out **C**.

CLUES:

1) **4 is not a factor of C.**
2) **C $\leq$ 29**
3) **C $>$ 21**
4) **3 is a factor of C.**

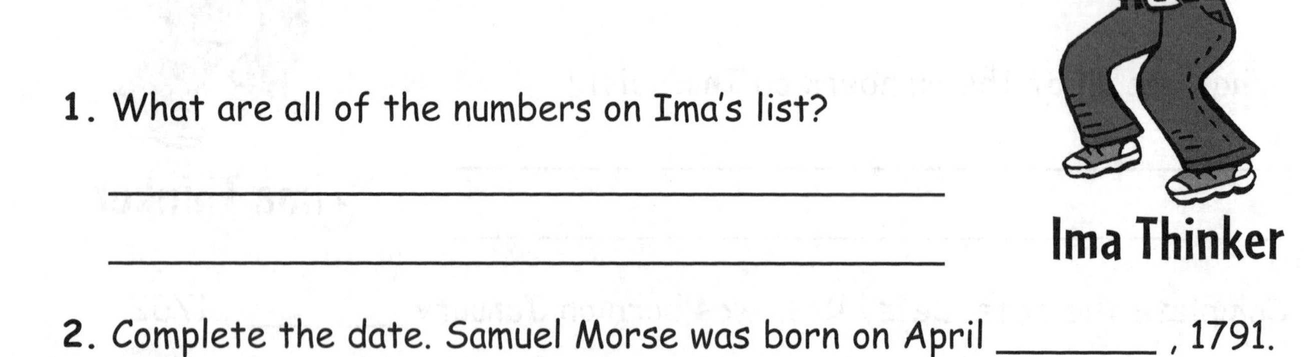

1. What are all of the numbers on Ima's list?

____

____

2. Complete the date. Samuel Morse was born on April ________ , 1791.

3. How did you figure out the birth date? ______________________________

____

4. Check your number with the clues. Show your work here.

Name ______________________ Date ______________

BIRTHDAY BOGGLE

PROBLEM 3

Benjamin Franklin was a scientist, an inventor, and a politician. He invented the lightning rod, bifocal glasses, and the odometer. The odometer is used to keep track of distances. Benjamin Franklin was born on January **G**, 1706.

The letter **G** stands for the date that Benjamin Franklin was born.

Use the clues to figure out **G**.

CLUES:

1) **G** $\leq 2 \times 10$

2) **The product of the tens and ones digits of G is less than 8.**

3) **G** $\neq 9 + 11$

4) **The sum of the digits of G is not 7.**

5) **G** $\geq 4 \times 4$

I'll start with Clues **1** and **5** and make a list of numbers for **G**. The first three numbers are 16, 17, and 18.

Ima Thinker

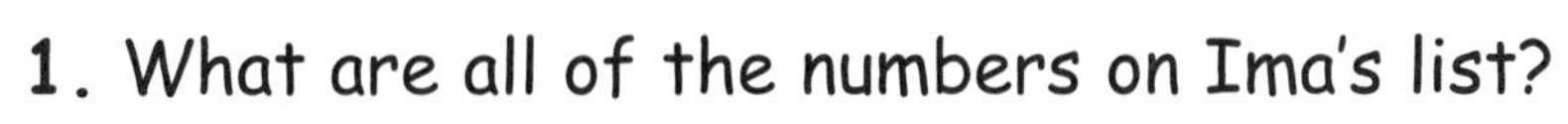
1. What are all of the numbers on Ima's list?

____

____

2. Complete the date. Benjamin Franklin was born on January ________ , 1706.

3. How did you figure out the birth date? ______________________________

____

4. Check your number with the clues. Show your work here.

Name ______________________________ Date ______________

PROBLEM 4

BIRTHDAY BOGGLE

Abraham Lincoln was the 16th president of the United States. He is most famous for his Emancipation Proclamation, which freed all slaves in the United States. Abraham Lincoln was born on February **S**, 1809.

The letter **S** stands for the date that Abraham Lincoln was born.

Use the clues to figure out **S**.

CLUES:

1) $S > 18 \div 2$
2) 4 is a factor of **S**.
3) 3 is a factor of **S**.
4) $S + S \leq 6 \times 8$
5) The difference between the two digits of **S** is less than 2.

1. Complete the date. Abraham Lincoln was born on February _____ , 1809.
2. How did you figure out the birth date? ______________________________

____

3. Check your number with the clues. Show your work here.

4. Abraham Lincoln signed the Emancipation Proclamation in the year he had his 54th birthday. In what year did he sign the Emancipation Proclamation? __________

Name ______________________________ Date ______________

BIRTHDAY BOGGLE

PROBLEM 5

Amelia Earhart was an aviator and the first woman to fly solo across the Atlantic Ocean. Amelia Earhart was born on July __F__, 1897.

The letter **F** stands for the date that Amelia Earhart was born.

Use the clues to figure out **F**.

CLUES:

1) **The difference between the two digits of F is not 3.**
2) **$F \times F \geq 400$**
3) **2 is a factor of F.**
4) **3 is a factor of F.**
5) **$2 \times F \leq 60$**

1. Complete the date. Amelia Earhart was born on July _____ , 1897.

2. How did you figure out the birth date? ______________________________

____

3. Check your number with the clues. Show your work here.

4. Amelia Earhart flew across the Atlantic in the year she had her 35th birthday.

In what year did she fly across the Atlantic? __________

Name ______________________________ Date ______________

PROBLEM 6

BIRTHDAY BOGGLE

John F. Kennedy was the 35th president of the United States. He also wrote the book *Profiles in Courage*. John F. Kennedy was born on May __**B**__, 1917.

The letter **B** stands for the date that John F. Kennedy was born.

Use the clues to figure out **B**.

CLUES:

1) **B** + **B** > 50
2) **B** ≠ 23 + 4
3) 2 is not a factor of **B**.
4) **B** ≤ 36
5) The product of the two digits of **B** is an even number.

1. Complete the date. John F. Kennedy was born on May ______ , 1917.

2. How did you figure out the birth date? ______________________________

____

3. Check your number with the clues. Show your work here.

4. John F. Kennedy was elected president of the United States when he was 43 years old. In what year was he elected president? __________

Name ______________________ Date ____________

PROBLEM **7**

BIRTHDAY BOGGLE

George Herman Ruth, better known as Babe Ruth, was a great baseball player. He was born on February **Q**, 1895.

The letter **Q** stands for the date that Babe Ruth was born.

Use the clues to figure out **Q**.

CLUES:

1) 3 is a factor of **Q**.
2) $Q \neq 8 + 4$
3) $Q \times Q \geq 5 \times 5$
4) $Q < 3 \times 6$
5) 2 is a factor of **Q**.

1. Complete the date. Babe Ruth was born on February _____ , 1895.

2. How did you figure out the birth date? ______________________

____

3. Check your number with the clues. Show your work here.

4. Babe Ruth was named the Greatest Player Ever 74 years after he was born. In what year did Babe receive this honor? ________

Menu Matters

Overview

Presented with signs of special menu deals, each showing the total cost of two categories of items, students solve for the cost of each item. This is preparation for solving systems of equations with one or two unknowns.

Algebra Focus

Solve two equations with two unknowns • Replace unknowns with their values

Problem-Solving Strategies

Reason deductively • Test cases

Related Math Skills ≤ ≥ x ÷

Compute with amounts of money

Math Language

Replace • Sum • Total cost

Introducing the Problem Set

Make photocopies of "Solve the Problem: Menu Matters" (page 44) and distribute to students. Have students work in pairs, encouraging them to discuss strategies they might use to solve the problem. You may want to walk around and listen in on some of their discussions. After a few minutes, display the problem on the board (or on the overhead if you made a transparency) and use the following questions to guide a whole-class discussion on how to solve the problem:

- What is in Special #1? *(2 chicken sandwiches and 2 ears of corn for $16.00.)*
- What is in Special #2? *(3 chicken sandwiches and a $4.00 quart of lemonade for a total of $22.00.)*
- What is the total cost of 3 chicken sandwiches without the lemonade? *($22.00 – $4.00, or $18.00.)*

Name ______ Date ______

MENU MATTERS

SOLVE THE PROBLEM

How much is a chicken sandwich?

Ben's Brunches has 2 specials.

The signs are clues to the costs of the items.

1. How did Ima figure out the cost of a chicken sandwich? ______

2. How can you figure out the cost of an ear of corn? ______

3. Replace each item in the signs with its cost. Check. Do the sums match the total costs? ______

44

- How can you figure out the cost of the corn? *(Replace each chicken sandwich in Special #1 with its cost. Then [2 x $6.00]+ 2 ears of corn is $16.00. So, the cost of the two ears of corn is $16.00 – $12.00, or $4.00, and each ear of corn is $4.00 ÷ 2, or $2.00.)*

Math Chat With the Transparency

Display the "Make the Case: Menu Matters" transparency on the overhead. Before students can decide which character is "on the ball," they need to figure out the answer to the problem. Encourage students to work in pairs to solve the problem, then bring the class together for another whole-class discussion. Ask:

- Who has the right answer? *(Buddy)*
- How did you figure it out? *(In Special #1, 2 tuna sandwiches without the bag of chips are $12.00 – $2.00, or $10.00, and each tuna sandwich is $10.00 ÷ 2, or $5.00. In Special #2, replace the tuna sandwich with its cost. Then the egg salad sandwich is $9.00 – $5.00, or $4.00.)*
- How do you think Sally got the answer of $5.00? *(She might have mistakenly given the cost for the tuna sandwich.)*
- How do you think that Bobby got the answer of $4.50? *(In Special #2, he probably thought that both sandwiches had the same cost. He then divided $9.00 by 2 and got $4.50.)*

Name ______ Date ______

MENU MATTERS

MAKE THE CASE

How much is an egg salad sandwich?

Carl's Cafe has 2 specials.

The signs are clues to the costs of the items.

Who is on the ball?

45

Name ______________________ Date ____________

SOLVE THE PROBLEM

How much is a chicken sandwich?

Ben's Brunches has 2 specials.

The signs are clues to the costs of the items.

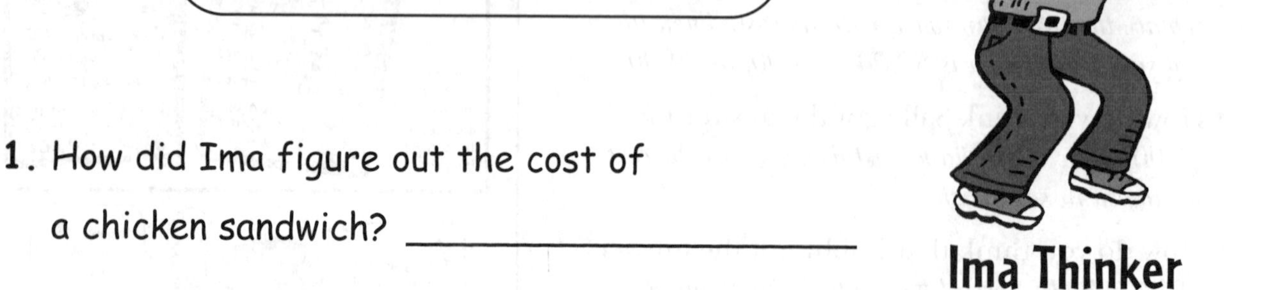

1. How did Ima figure out the cost of a chicken sandwich? ______________________

____

2. How can you figure out the cost of an ear of corn?

____

3. Replace each item in the signs with its cost. Check.

   Do the sums match the total costs? ____________

Name ____________________ Date ____________

MENU MATTERS

MAKE THE CASE

How much is an egg salad sandwich?

Carl's Cafe
Combo Special #1
2 tuna sandwiches
and
1 $2.00 bag of chips
Today's Price
$12.00

Carl's Cafe has 2 specials. The signs are clues to the costs of the items.

That's easy. The egg salad sandwich is $5.00.

No way. The egg salad sandwich is $4.00.

Sally Soccer

Obviously the egg salad sandwich is $4.50.

Bobby Baseball

Who is on the ball?

Name ______________________ Date __________

PROBLEM 1

MENU MATTERS

How much is a banana smoothie?

Rory's Restaurant has 2 specials.

The signs are clues to the costs of the items.

I'll start with Special #1. I can figure out the cost of one banana smoothie.

Ima Thinker

1. How did Ima figure out the cost of a banana smoothie?

____

2. How can you figure out the cost of an orange smoothie?

____

3. Replace each item in the signs with its cost. Check.

   Do the sums match the total costs? __________

Name ____________________ Date ____________

MENU MATTERS

PROBLEM 2

How much is a peanut butter and jelly sandwich?

Easy Eats has 2 specials.

The signs are clues to the costs of the items.

I'll start with Special #2. Without the cookies, the total cost of the 2 peanut butter and jelly sandwiches is \$11.00 – \$3.00, or \$8.00.

Ima Thinker

1. How did Ima figure out the cost of a peanut butter and jelly sandwich? ____________________

2. How can you figure out the cost of a dish of ice cream?

____

3. Replace each item in the signs with its cost. Check.

   Do the sums match the total costs? ____________

Name ______________________________ Date ______________

PROBLEM 3

How much is an onion roll?

Dale's Diner has 2 specials.

The signs are clues to the costs of the items.

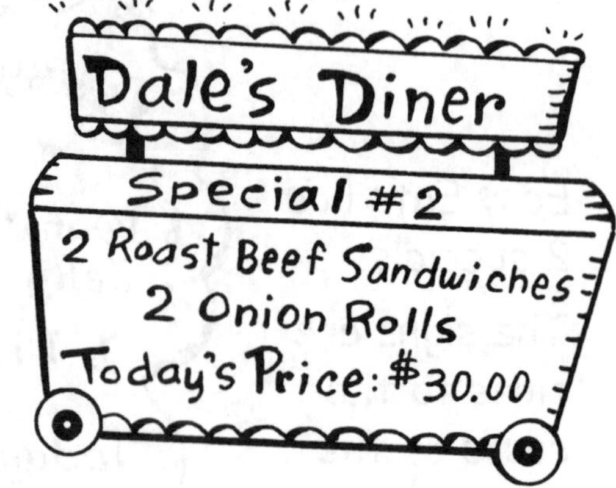

I'll start with Special #1. Without the bottle of water, the total cost of the 4 onion rolls is $27.00 – $3.00, or $24.00.

Ima Thinker

1. How did Ima figure out the cost of an onion roll?

____

2. How can you figure out the cost of a roast beef sandwich?

____

3. Replace each item in the signs with its cost. Check.

   Do the sums match the total costs? __________

Name ______________________________ Date ______________

MENU MATTERS

PROBLEM 4

How much is a hamburger?

Dina's Dinette has 2 specials.

The signs are clues to the costs of the items.

1. How can you figure out the cost of a hamburger?

____

2. How can you figure out the cost of a club sandwich?

____

3. Replace each item in the signs with its cost. Check.

   Do the sums match the total costs? __________

Name ______________________________ Date ______________

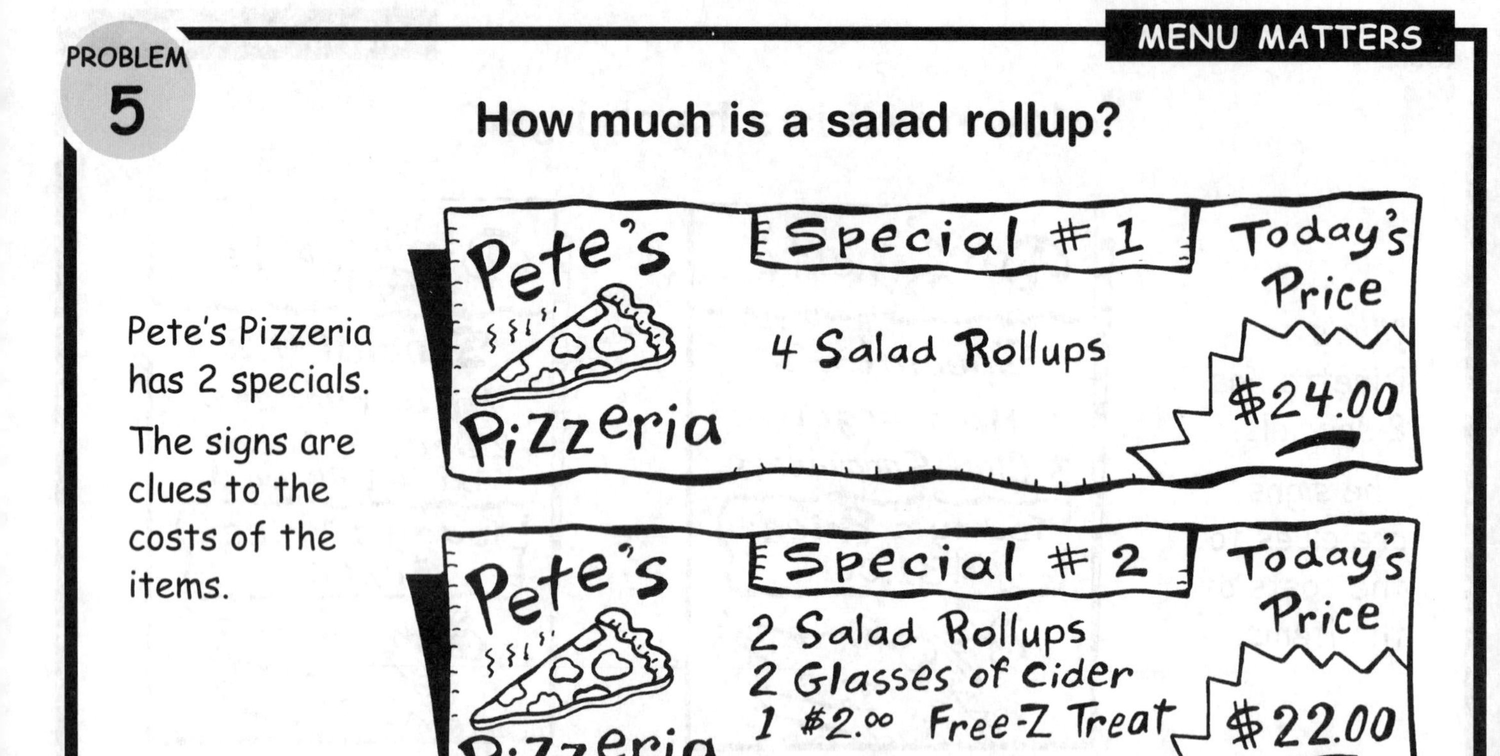

1. How much is a salad rollup? ____________

2. How can you figure out the cost of a glass of cider?

____

3. Replace each item in the signs with its cost. Check.

   Do the sums match the total costs? ____________

Name ______________________ Date ______________

MENU MATTERS

PROBLEM 6

How much is a hot dog?

Baker's Best has 2 specials.

The signs are clues to the costs of the items.

1. How much is a hot dog? ____________

2. How much is a baked potato? ____________

3. How did you figure out the cost of each item? ____________________

____

Name ______________________ Date ____________

PROBLEM 7

MENU MATTERS

How much is an egg?

Benny's Breakfast has 2 specials.

The signs are clues to the costs of the items.

1. How much is an egg? ____________
2. How much is a pancake? ____________
3. How did you figure out the cost of each item? ____________

____

ABC Code Crackers

Overview

Presented with a grid of letters that represent numbers, students use the column sums to figure out the value of each letter.

Algebra Focus

Solve equations with two or three unknowns • Replace unknowns with their values • Recognize that same symbols have the same value • Understand that taking away an addend changes the sum by the same amount

Problem-Solving Strategies ????????

Reason deductively • Test cases

Related Math Skills ≤ ≥ X ÷

Compute with whole numbers

Math Language

Replace • Symbol • Value

Introducing the Problem Set

Make photocopies of "Solve the Problem: ABC Code Crackers" (page 55) and distribute to students. Have students work in pairs, encouraging them to discuss strategies they might use to solve the problem. You may want to walk around and listen in on some of their discussions. After a few minutes, display the problem on the board (or on the overhead if you made a transparency) and use the following questions to guide a whole-class discussion on how to solve the problem:

- Look at the columns. How are they alike? *(They all have numbers at the top of the columns. There are three letters or numbers in each column. All of the columns contain at least one A.)*
- What are the numbers at the tops of the columns? *(Sums of the numbers and the values of the letters in the columns.)*

- What is in the first column? *(A, B, and C)* The second column? *(6, B, and A)* The third column? *(C, C, and A)*
- Why did Ima start with the second column? *(She can figure out that A + B = 19 – 6, or 13. Then she can replace A and B in the first column with 13 and figure out the value of C. Once she knows C, she can figure out the value of A in the third column, and then the value of B.)*
- Since A + B is 13, what is the value of C? *(20 – 13, or 7)*
- How can you figure out the value of A? *(In the third column 7 + 7 + A = 23, so A = 9.)*
- How can you figure out the value of B? *(Replace A with 9 in the second column and solve for B: 19 – 6 – 9 = 4; or replace A with 9 and C with 7 in the first column, and solve for B: 20 – 9 – 7 = 4.*

Work together as a class to answer the questions in "Solve the Problem: ABC Code Crackers."

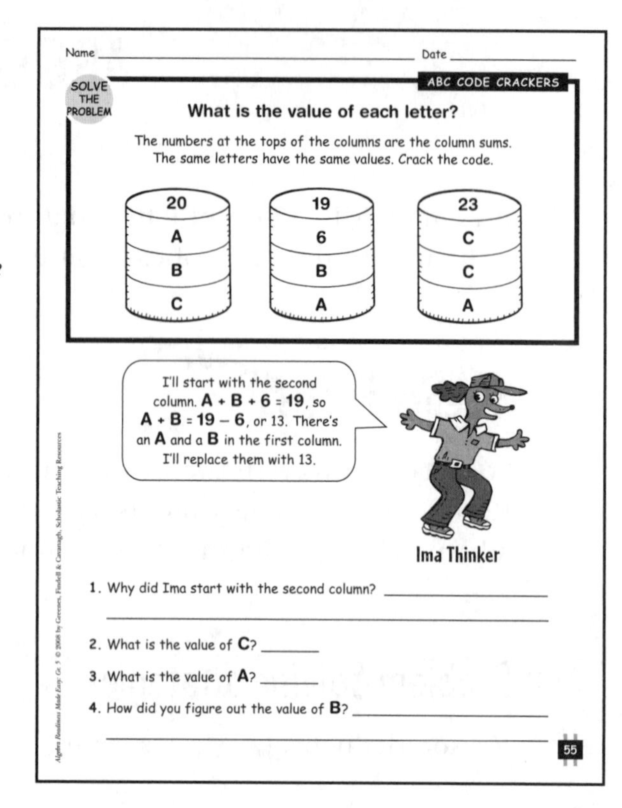

Name ______________________ Date ____________

SOLVE THE PROBLEM

ABC CODE CRACKERS

What is the value of each letter?

The numbers at the tops of the columns are the column sums. The same letters have the same values. Crack the code.

20	19	23
A	6	C
B	B	C
C	A	A

1. Why did Ima start with the second column? ______________________
2. What is the value of **C**? ________
3. What is the value of **A**? ________
4. How did you figure out the value of **B**? ______________________

55

Math Chat With the Transparency

Display the "Make the Case: ABC Code Crackers" transparency on the overhead. Before students can decide which character is "on the ball," they need to figure out the answer to the problem. Encourage students to work in pairs to solve the problem, then bring the class together for another whole-class discussion. Ask:

- Who has the right answer? *(Buddy)*
- How did you figure it out? *(In the first column, A + B = 13 – 2, or 11. In the second column, replace A and B with 11. Then A = 16 – 11, or 5. In the first column, replace A with 5. Then B = 13 – 5 – 2, or 6. In the third column, replace B with 6. Then C + C = 8 – 6, or 2, and each C is 2 ÷ 2, or 1)*

Name ______________________ Date ____________

MAKE THE CASE

ABC CODE CRACKERS

What is the value of C?

The numbers at the tops of the columns are the column sums. The same letters have the same values. Crack the code.

13	16	8
A	A	C
2	B	C
B	A	B

56

Who is on the ball?

- How do you think Sally got the answer of 6? *(She probably solved for B instead of for C.)*
- How do you think Bobby got the answer of 5? *(He probably solved for A instead of for C.)*

Name ______________________ Date ____________

ABC CODE CRACKERS

SOLVE THE PROBLEM

What is the value of each letter?

The numbers at the tops of the columns are the column sums.
The same letters have the same values. Crack the code.

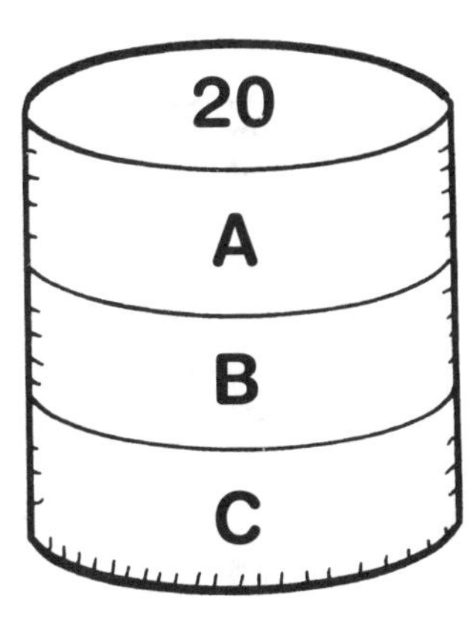

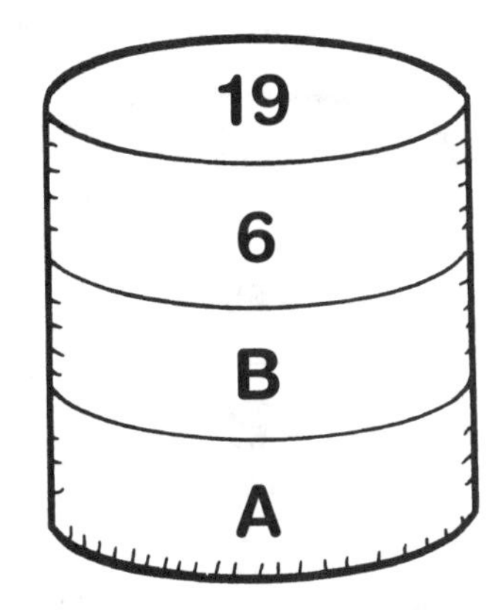

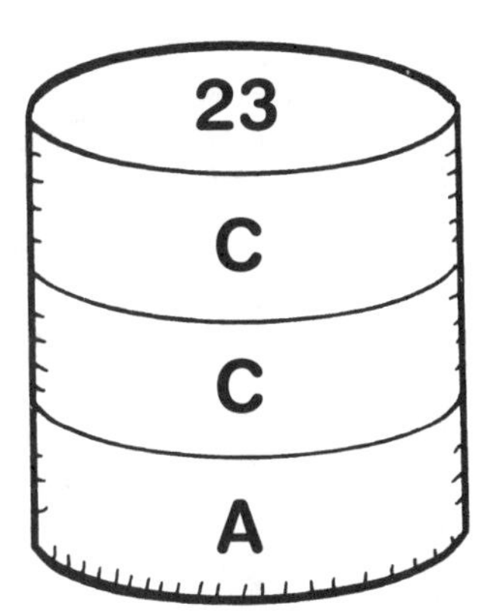

I'll start with the second column. **A** + **B** + **6** = **19**, so **A** + **B** = **19** – **6**, or 13. There's an **A** and a **B** in the first column. I'll replace them with 13.

Ima Thinker

1. Why did Ima start with the second column? ______________________

____

2. What is the value of **C**? ________

3. What is the value of **A**? ________

4. How did you figure out the value of **B**? ______________________

____

Name ______________________ Date ______________

MAKE THE CASE

ABC CODE CRACKERS

What is the value of C?

The numbers at the tops of the columns are the column sums.
The same letters have the same values. Crack the code.

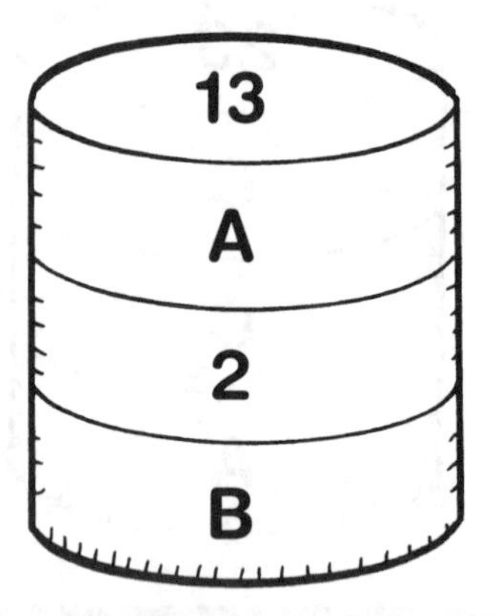

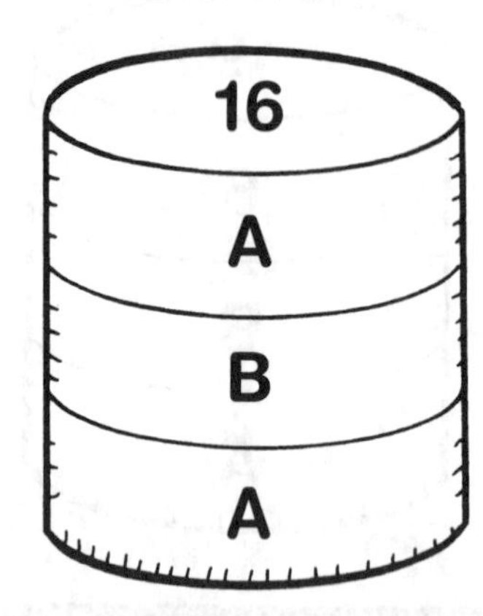

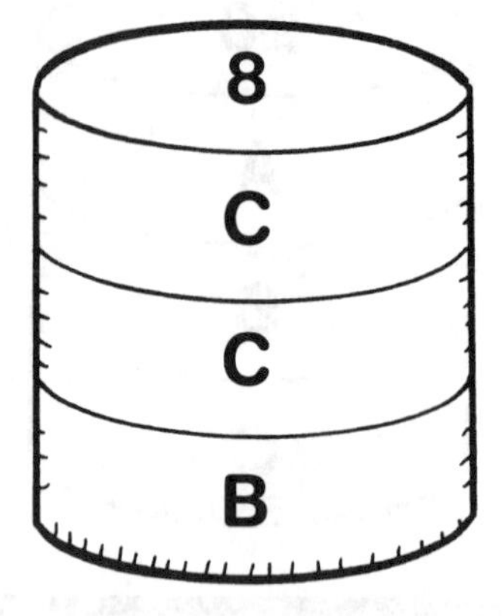

After studying the problem, I am sure that **C** is 6.

Buddy Basketball

C is no doubt equal to 1.

Sally Soccer

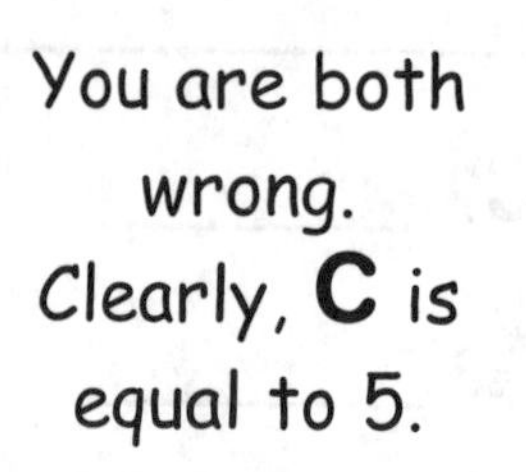

You are both wrong. Clearly, **C** is equal to 5.

Bobby Baseball

Who is on the ball?

Name ______________________ Date ____________

PROBLEM 1

What is the value of each letter?

The numbers at the tops of the columns are the column sums.
The same letters have the same values. Crack the code.

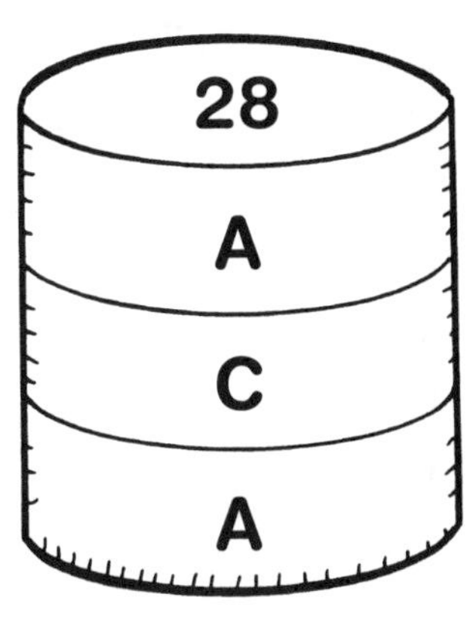

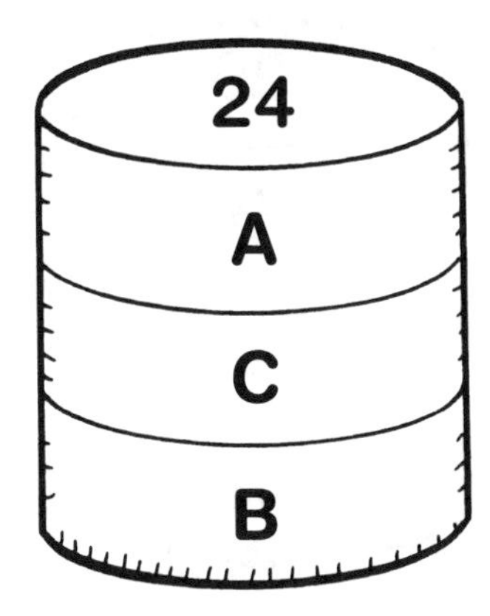

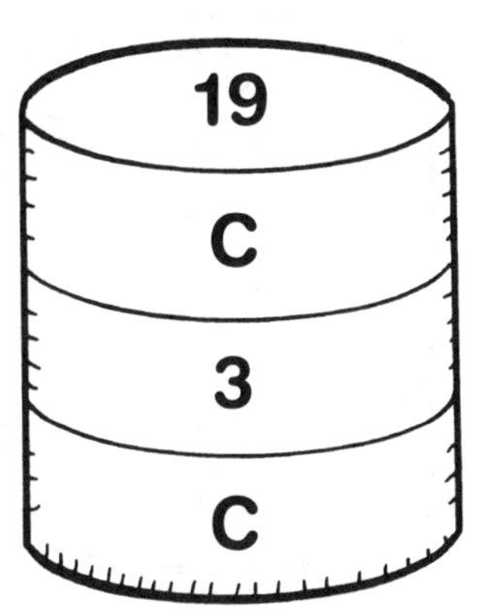

I'll start with the third column.
C + 3 + C = 19,
so **C + C = 19 – 3**, or 16.
C is 16 ÷ 2, or 8. In the first column I'll replace **C** with 8.

Ima Thinker

1. Why did Ima start with the third column? ______________________

____

2. What is the value of **A**? __________

3. How did you figure out the value of **B**? ______________________

____

4. What is the value of **A + A + B + B + C + C**? __________

Name ______________________ Date __________

PROBLEM 2

ABC CODE CRACKERS

What is the value of each letter?

The numbers at the tops of the columns are the column sums.
The same letters have the same values. Crack the code.

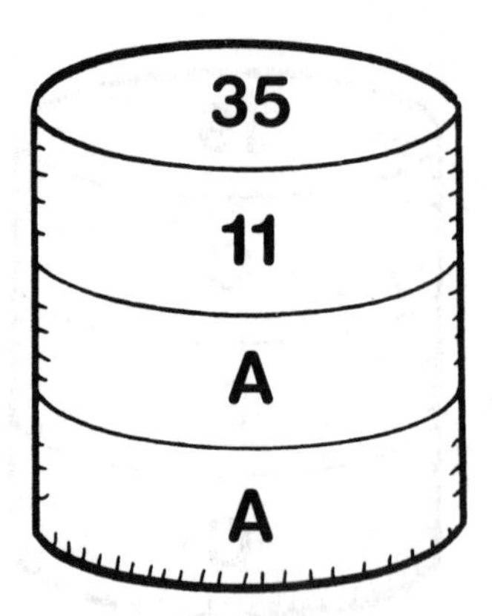

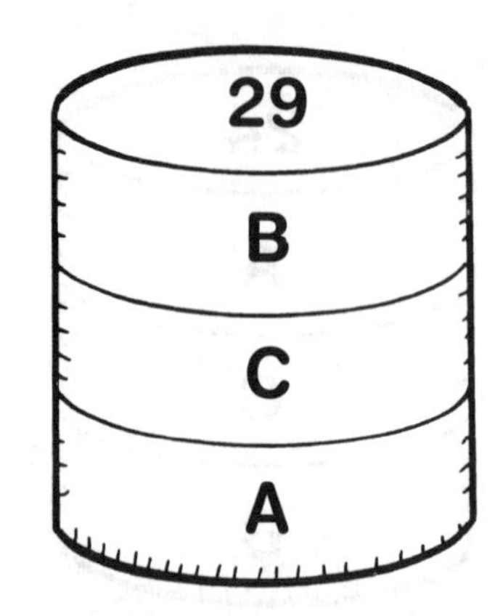

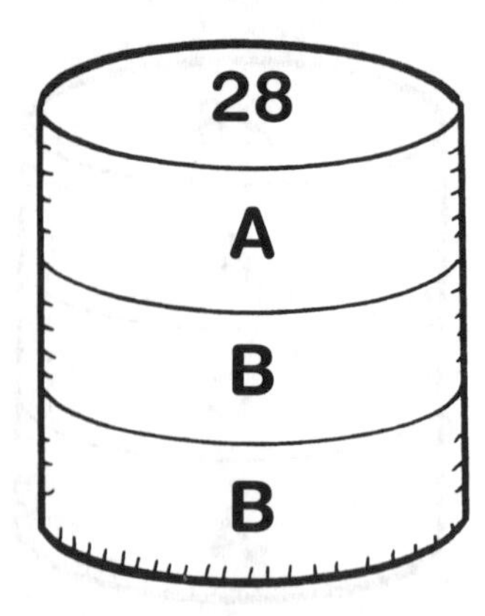

I'll start with the first column.
11 + A + A = 35, so
A + A = 35 – 11, or 24.
A is 24 ÷ 2, or 12. In the third column, I'll replace **A** with 12.

Ima Thinker

1. Why did Ima start with the first column? ______________________

____

2. What is the value of **B**? __________

3. How did you figure out the value of **C**? ______________________

____

4. What is the value of (2 x **A**) + **B** + **C**? __________

Name ________________________________ Date ________________

PROBLEM 3

ABC CODE CRACKERS

What is the value of each letter?

The numbers at the tops of the columns are the column sums.
The same letters have the same values. Crack the code.

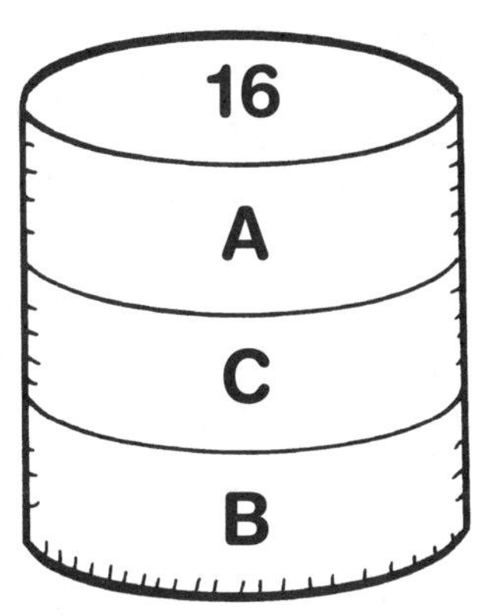

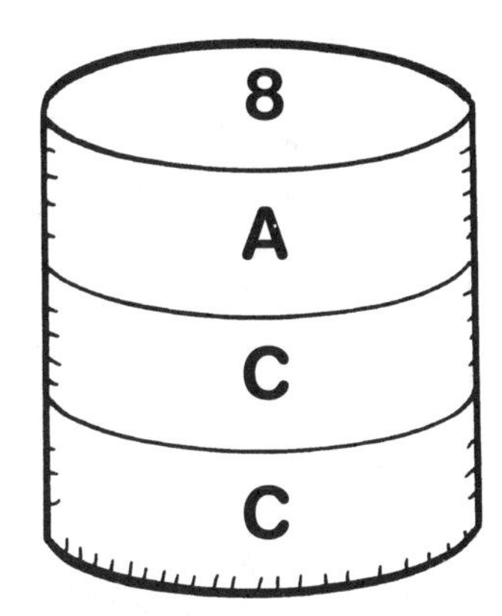

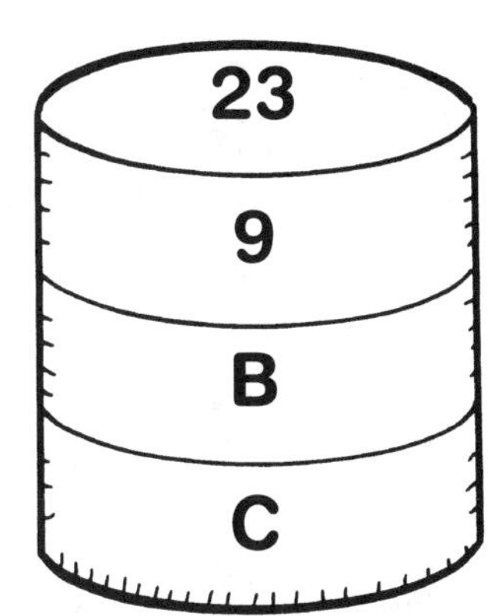

I'll start with the third column. **B** + **C** = 23 – 9, or 14. **B** and **C** are in the first column. I'll replace them with 14.

Ima Thinker

1. Why did Ima start with the third column? ______________________

____

2. What is the value of **A**? ________

3. How did you figure out the value of **A**? ______________________

____

4. What is the value of (3 × **A**) + (2 × **B**) + **C**? ________

Name ______________________________ Date ______________

PROBLEM 4

ABC CODE CRACKERS

What is the value of each letter?

The numbers at the tops of the columns are the column sums.
The same letters have the same values. Crack the code.

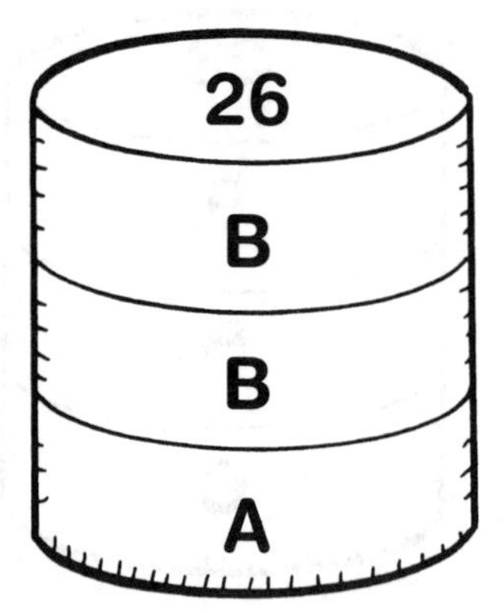

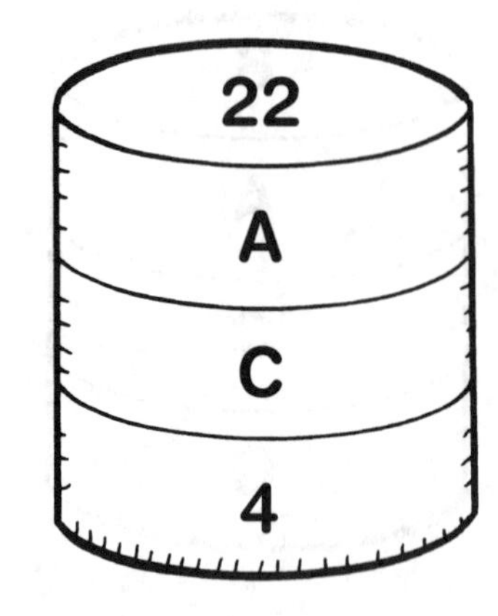

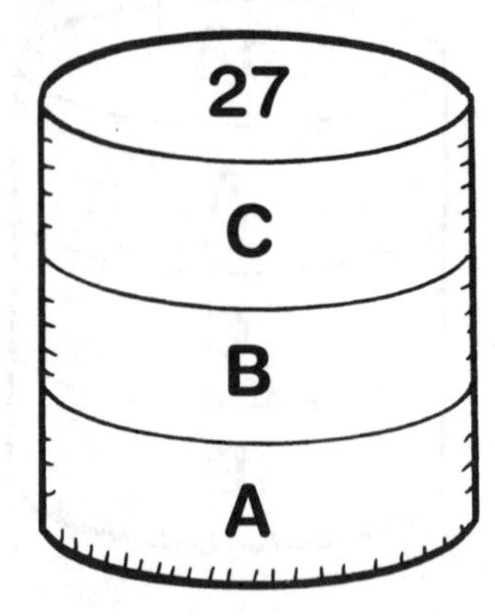

1. What is the value of **A** + **C**? ________

2. What is value of **B**? ________

3. How did you figure out the value of **C**? ______________________________

____

4. What is the value of (2 × **A**) + **B** – **C**? ________

Name ______________________ Date ____________

PROBLEM 5

ABC CODE CRACKERS

What is the value of each letter?

The numbers at the tops of the columns are the column sums.
The same letters have the same values. Crack the code.

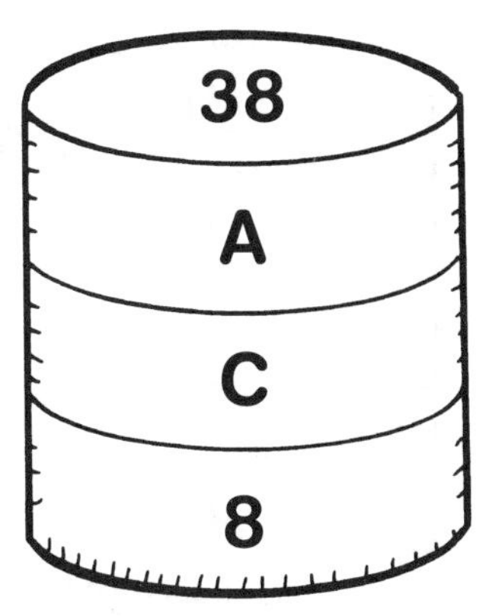

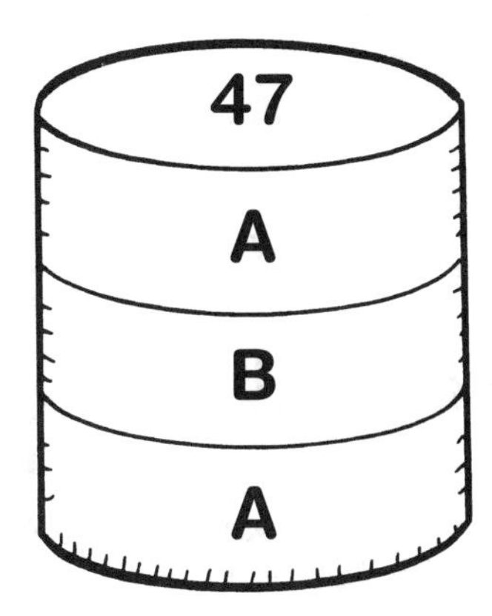

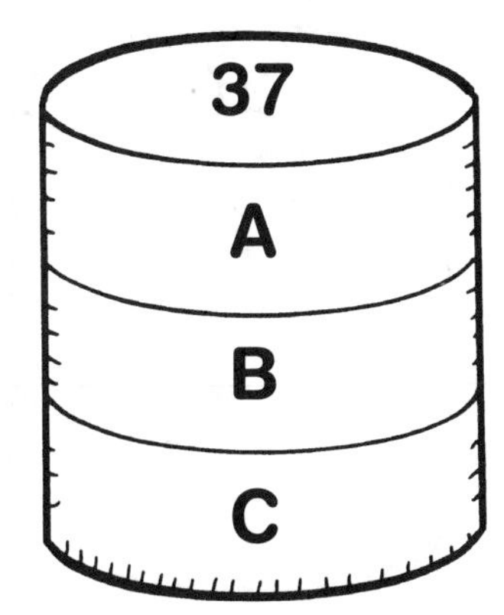

1. What is the value of **A** + **C**? ________

2. What is value of **B**? ________

3. What is value of **A**? ________

4. How did you figure out the value of **C**? ______________________________

____

Name ______________________ Date ______________

PROBLEM 6

ABC CODE CRACKERS

What is the value of each letter?

The numbers at the tops of the columns are the column sums.
The same letters have the same values. Crack the code.

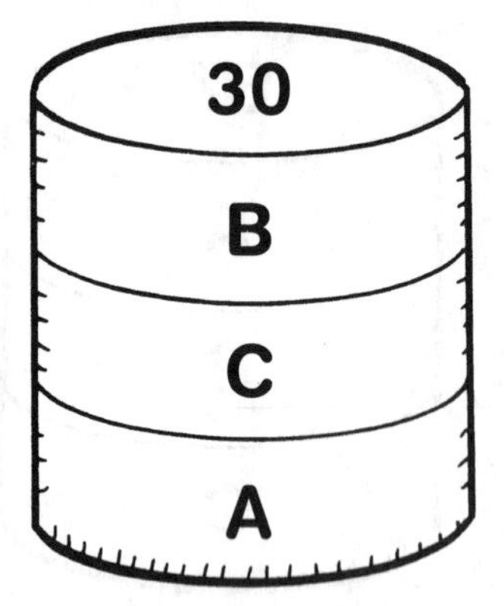

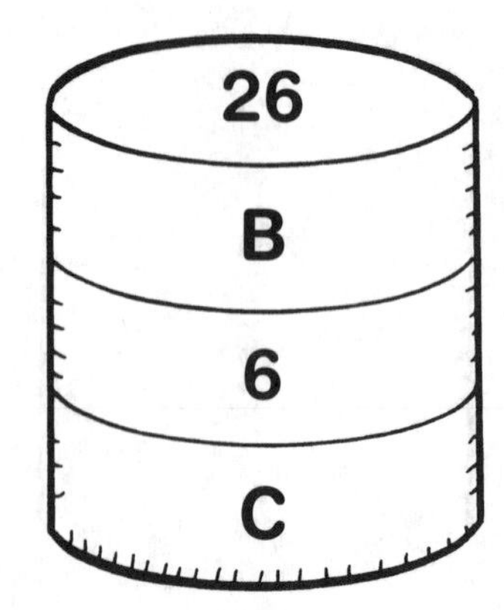

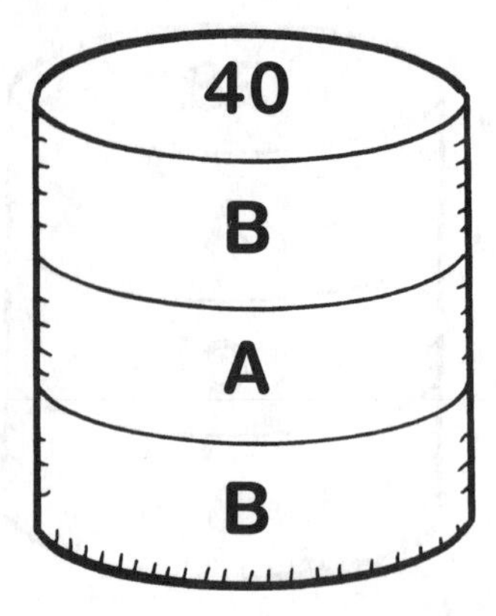

1. What is the value of **B** + **C**? ________
2. What is value of **A**? ________
3. What is value of **B**? ________
4. How did you figure out the value of **C**? ______________________

____

Name ______________________________ Date ______________

PROBLEM 7

ABC CODE CRACKERS

What is the value of each letter?

The numbers at the tops of the columns are the column sums.
The same letters have the same values. Crack the code.

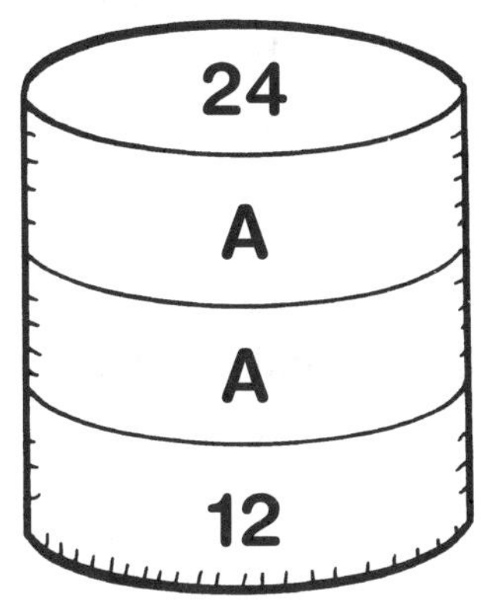

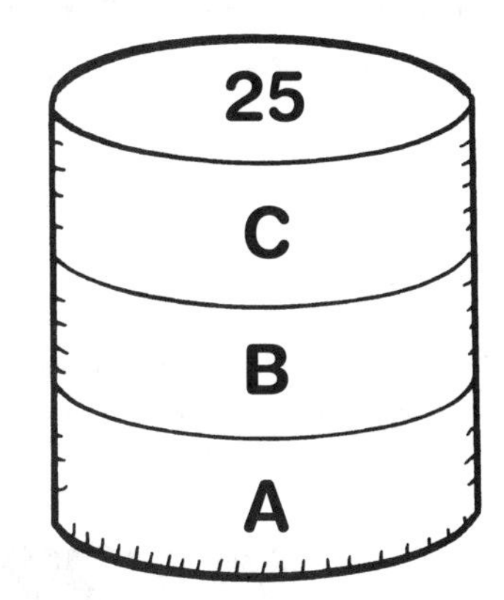

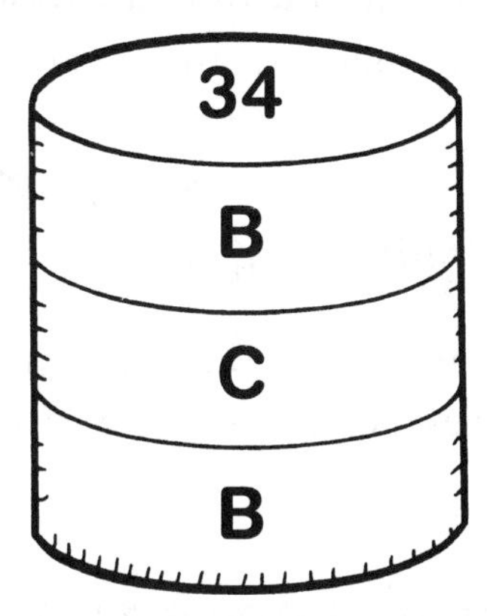

1. What is value of **A**? ________
2. What is the value of **B** + **C**? ________
3. What is value of **B**? ________
4. How did you figure out the value of **C**? ______________________________

____

Balance the Blocks

Overview

Presented with clues about the relative weights of three different types of blocks in a pan balance, students figure out which blocks will balance a new set of blocks.

Algebra Focus

Understand that substituting one set of blocks with a second set of equal weight preserves balance • Explore the concept of equality

Problem-Solving Strategies

Reason about proportional relationships • Reason deductively

Related Math Skills ≤ ≥ X ÷

Compute with whole numbers

Math Language

Balance • Cube • Cylinder • Sphere • Substitute • Weigh the same

Introducing the Problem Set

Make photocopies of "Solve the Problem: Balance the Blocks" (page 66) and distribute to students. Have students work in pairs, encouraging them to discuss strategies they might use to solve the problem. You may want to walk around and listen in on some of their discussions. After a few minutes, display the problem on the board (or on the overhead if you made a transparency) and use the following questions to guide a whole-class discussion on how to solve the problem:

- Look at the pan balances. What is in the first pan balance? *(4 spheres in the pan on the left balancing 2 cylinders in the pan on the right)* What is in the second pan balance? *(1 cylinder in the pan on the left balancing 3 cubes in the pan on the right)*

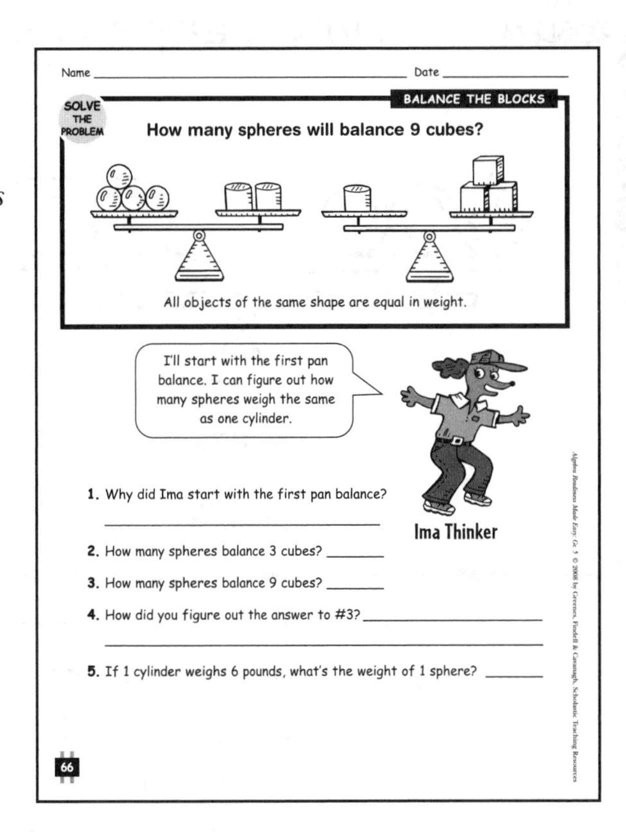

- Which weighs more, 1 sphere or 1 cylinder? *(1 cylinder)* How did you decide? *(It takes 4 spheres to balance 2 cylinders, so the cylinder weighs twice as much as a sphere, or a sphere is half the weight of 1 cylinder.)*
- What do you need to find out? *(How many spheres will balance 9 cubes.)*
- Why did Ima start with the first pan balance? *(To figure out how many spheres balance 1 cylinder.)*
- If 4 spheres balance 2 cylinders, how many spheres will balance one cylinder? *(2 spheres will balance 1 cylinder.)*

Work together as a class to answer the questions in "Solve a Problem: Balance the Blocks."

Math Chat With the Transparency

Display the "Make the Case: Balance the Blocks" transparency on the overhead. Before students can decide which character is "on the ball," they need to figure out the answer to the problem. Encourage students to work in pairs to solve the problem, then bring the class together for another whole-class discussion. Ask:

- Who has the right answer? *(Sally)*
- How did you figure it out? *(In the first pan balance, since 6 cubes balance 2 spheres, 3 cubes (6 ÷ 2) will balance 1 sphere (2 ÷ 2). In the second pan balance, substitute 3 cubes for 1 sphere. Since 3 cubes will balance 2 cylinders, then 9 cubes (3 x 3) will balance 6 cylinders (3 x 2).)*
- How do you think Buddy got the answer 18? *(He may have tripled the number of cylinders to get 6 cylinders, then tripled the number of cubes to get 18 cubes.)*
- How do you think Bobby got the answer 7? *(He may have decided that if 2 cylinders balance 3 cubes, then 6 cylinders (2 + 4) balance or 7 cubes (3 + 4).)*

Name ____________________ Date ____________

BALANCE THE BLOCKS

SOLVE THE PROBLEM

How many spheres will balance 9 cubes?

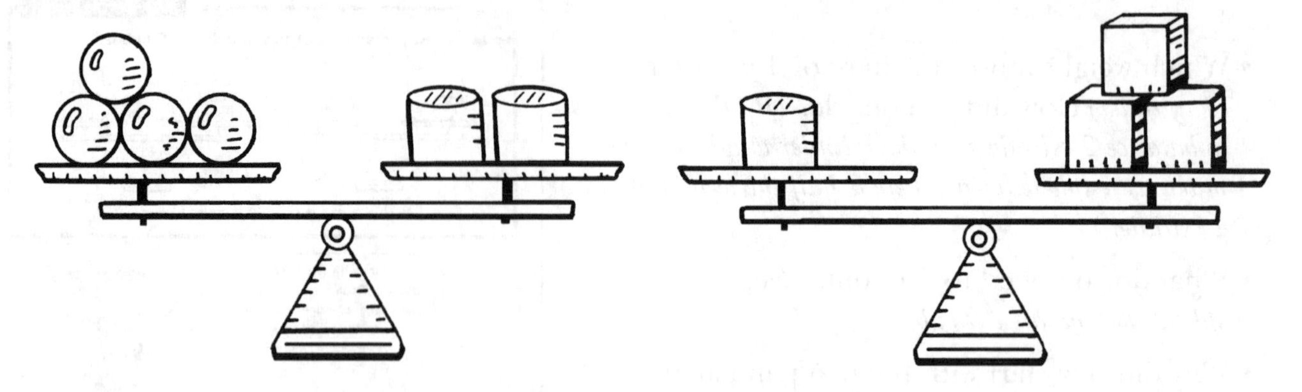

All objects of the same shape are equal in weight.

1. Why did Ima start with the first pan balance?

   ____

2. How many spheres balance 3 cubes? ________

3. How many spheres balance 9 cubes? ________

4. How did you figure out the answer to #3? ____________________

   ____

5. If 1 cylinder weighs 6 pounds, what's the weight of 1 sphere? ________

Name ____________________ Date ____________

MAKE THE CASE

BALANCE THE BLOCKS

How many cubes will balance 6 cylinders?

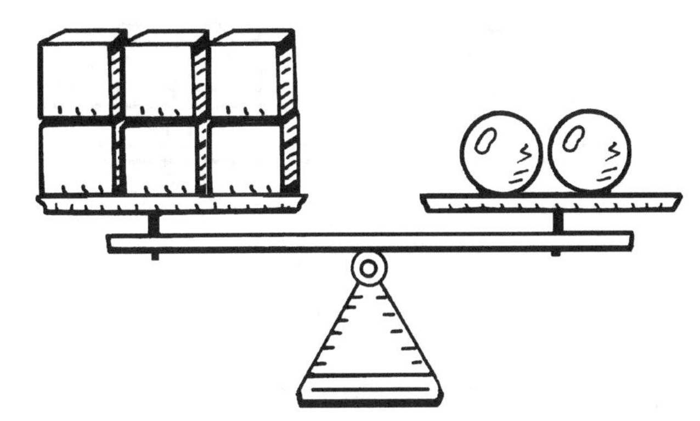

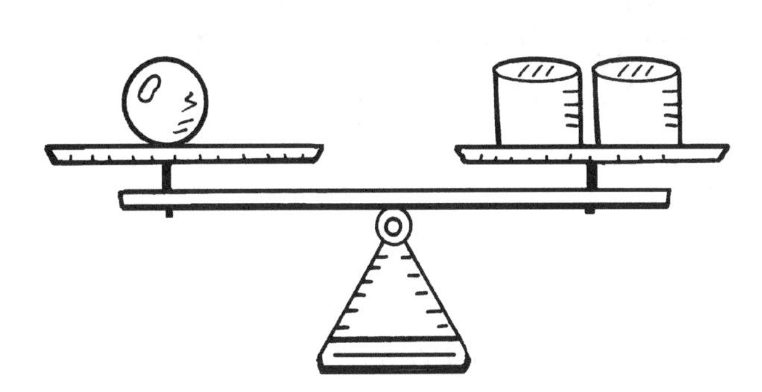

All objects of the same shape are equal in weight.

That's easy. Nine cubes will balance 6 cylinders.

No way. It's 18 cubes.

Buddy Basketball

Sally Soccer

You're both wrong. I am sure it's 7 cubes.

Bobby Baseball

Who is on the ball?

Name ______________________ Date ____________

PROBLEM 1

How many spheres will balance 4 cylinders?

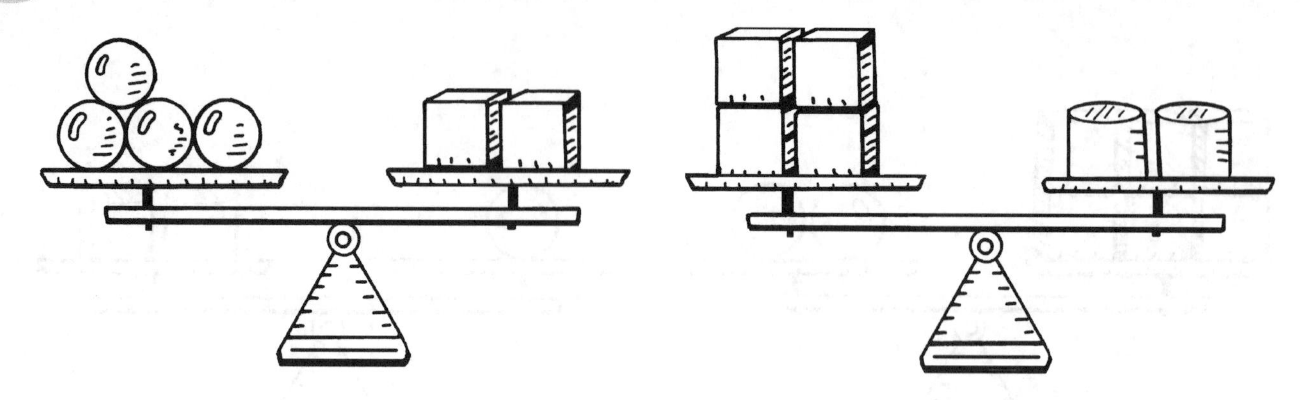

All objects of the same shape are equal in weight.

1. Why did Ima start with the first pan balance? ______________________

   ____

2. How many spheres balance 2 cylinders? ________

3. How many spheres balance 4 cylinders? ________

4. How did you figure out the answer to #3? ______________________

   ____

5. If 1 cube weighs 4 pounds, what's the weight of 1 cylinder? ________

Name ________________________________ Date ______________

BALANCE THE BLOCKS

PROBLEM 2

How many spheres will balance 4 cubes?

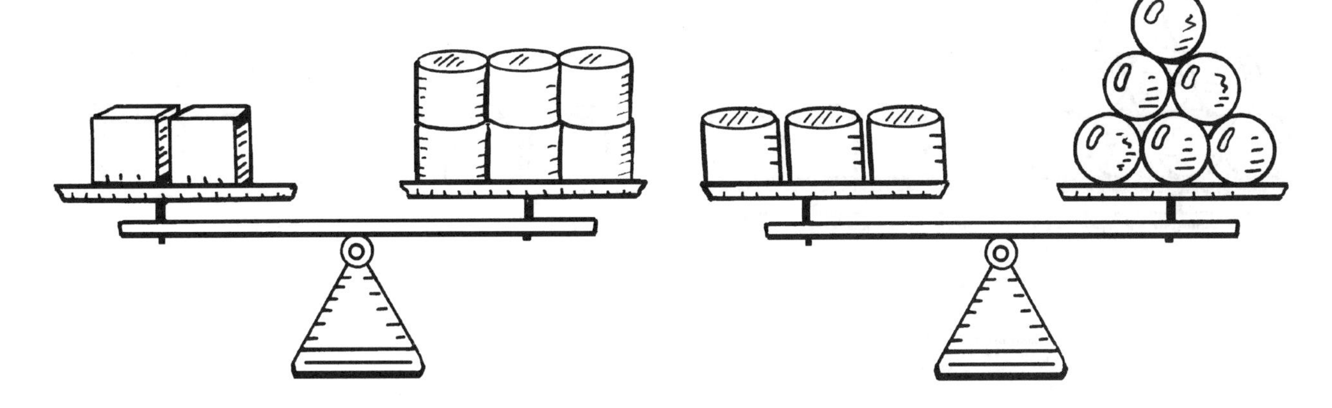

All objects of the same shape are equal in weight.

I'll start with the second pan balance. I can figure out how many spheres weigh the same as 1 cylinder.

Ima Thinker

1. Why did Ima start with the second pan balance? ______________________

____

2. How many spheres balance 2 cubes? ________

3. How many spheres balance 4 cubes? ________

4. How did you figure out the answer to #3? ______________________

____

5. If 1 cylinder weighs 2 pounds, what's the weight of 1 cube? __________

Name ______________________ Date ____________

PROBLEM 3

BALANCE THE BLOCKS

How many cubes will balance 6 spheres?

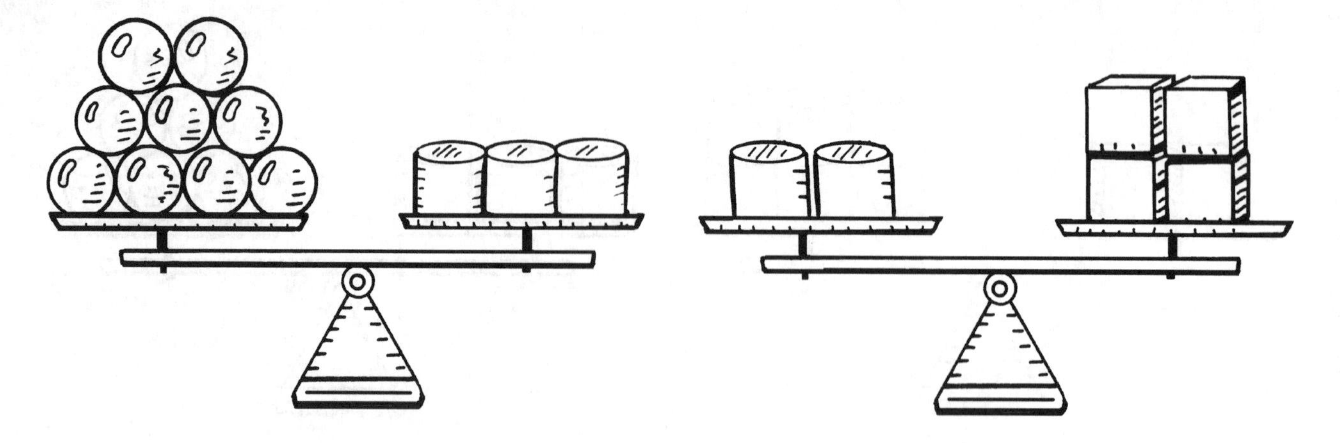

All objects of the same shape are equal in weight.

1. Why did Ima start with the second pan balance? ______________________

____

2. How many cubes will balance 9 spheres? ________

3. How many cubes will balance 6 spheres? ________

4. How did you figure out the answer to #3? ______________________

____

5. If 1 cylinder weighs 12 pounds, what's the weight of 1 sphere? ________

Name ______________________ Date __________

BALANCE THE BLOCKS

PROBLEM 4

How many cylinders will balance 9 spheres?

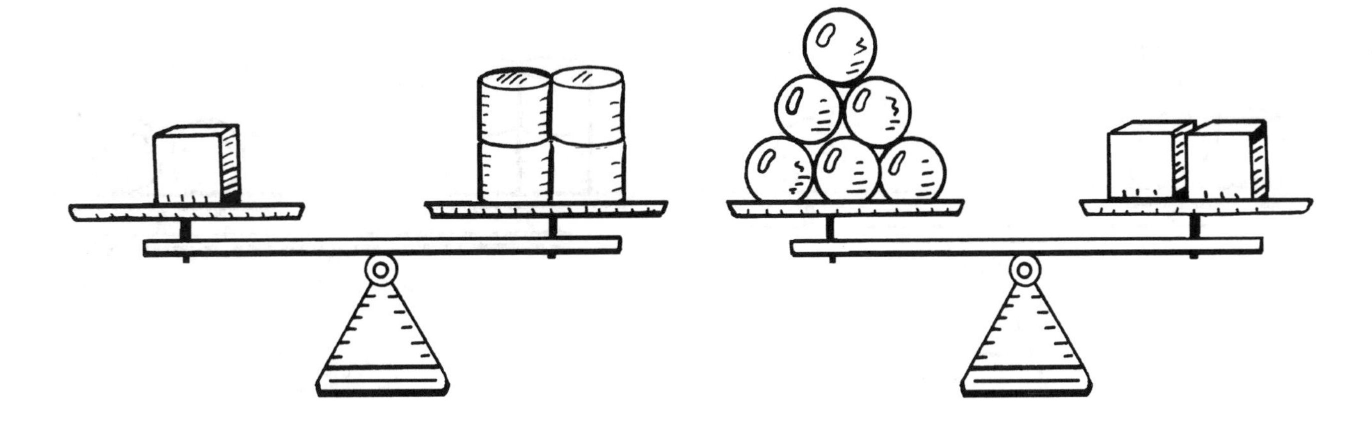

All objects of the same shape are equal in weight.

1. How many cubes will balance 3 spheres? ________

2. How many cylinders will balance 6 spheres? ________

3. How many cylinders will balance 9 spheres? ________

4. How did you figure out the answer to #3? ________________________

____

5. If 1 cube weighs 6 pounds, what's the weight of 1 sphere? __________

Name ____________________ Date ____________

PROBLEM 5

BALANCE THE BLOCKS

How many cubes will balance 6 cylinders?

All objects of the same shape are equal in weight.

1. How many cylinders will balance 1 sphere? ________
2. How many cubes will balance 4 cylinders? ________
3. How many cubes will balance 6 cylinders? ________
4. How did you figure out the answer to #3? ________________

   ____
5. If 2 spheres weigh 20 pounds, what's the weight of 1 cube? ________

Name ________________________________ Date ______________

PROBLEM 6

BALANCE THE BLOCKS

How many cylinders will balance 6 cubes?

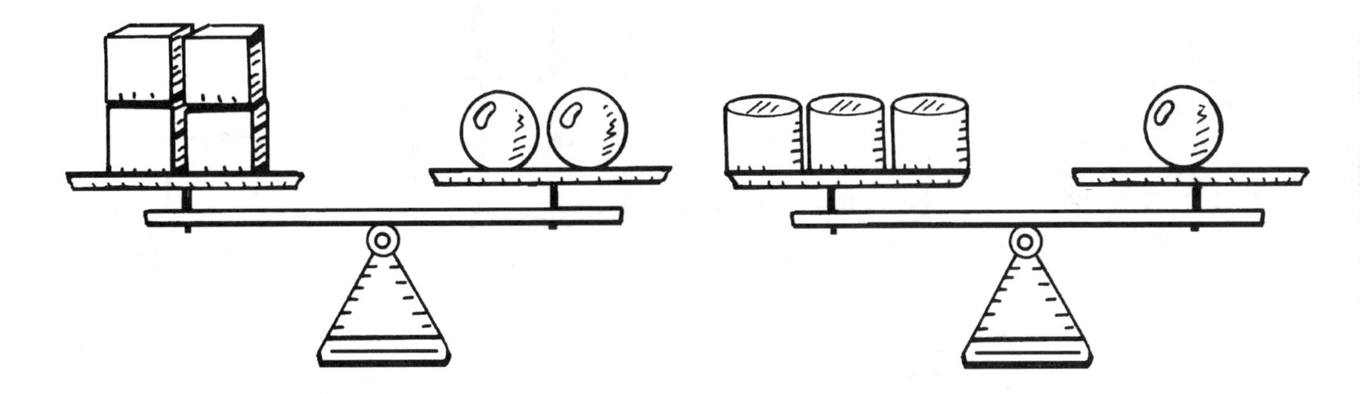

All objects of the same shape are equal in weight.

1. How many spheres will balance 2 cubes? ________

2. How many cylinders will balance 4 cubes? ________

3. How many cylinders will balance 6 cubes? ________

4. How did you figure out the answer to #3? ________________________

   ____

5. If 1 sphere weighs 8 pounds, what's the weight of 1 cube? ____________

Name ______________________ Date ____________

PROBLEM 7

BALANCE THE BLOCKS

How many cubes will balance 6 spheres?

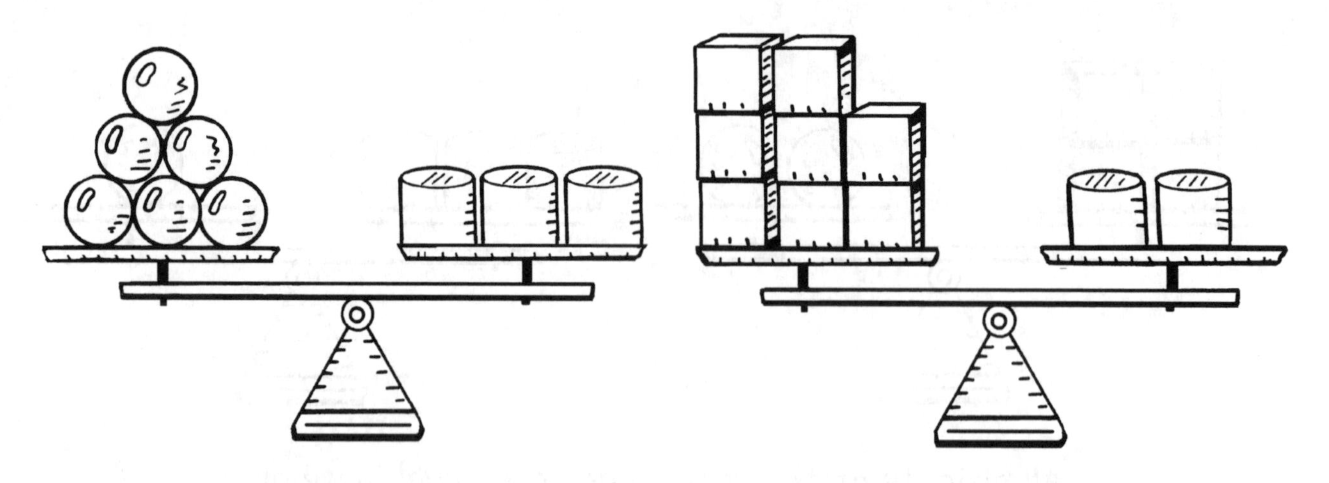

All objects of the same shape are equal in weight.

1. How many cubes will balance 1 cylinder? ________

2. How many cubes will balance 8 spheres? ________

3. How many cubes will balance 6 spheres? ________

4. How did you figure out the answer to #3? ________________

____

5. If 1 cylinder weighs 12 pounds, what's the weight of 1 cube? ________

SOLVE IT

1. **Look** What is the problem?

2. **Plan and Do** What will you do first? How will you solve the problem?

3. **Answer and Check** How can you be sure your answer is correct?

SOLVE IT: GRID PATTERNS

What is the greatest number in Sylvia's square?

1	2	3	4	5	6	7	8	9
10	11	12	13	14	15	16	17	18
19	20	21	22	23	24	25	26	27
28	29	30	31	32	33	34	35	36

The array of numbers continues. Sylvia drew a 3-by-3 square around 9 numbers in the array. The middle number in Sylvia's square is 56.

SOLVE IT: DOLLAR DILEMMA

What is the cost of Tiny's phone?

Use the facts to figure out the costs.

Tiny Sara Tiptoe Paul Frosty

FACTS:

A) Tiny's phone costs $\frac{1}{2}$ the total cost of Sara's and Tiptoe's phones.
B) Sara's phone costs \$6 more than Tiptoe's phone.
C) Tiptoe's phone costs \$18 plus $\frac{1}{2}$ the cost of Paul's phone.
D) Paul's phone costs \$30 plus $\frac{1}{3}$ the cost of Frosty's phone.
E) Frosty's phone costs \$42.

SOLVE IT: BIRTHDAY BOGGLE

Bruce Springsteen is a rock musician. He sings, writes music, and plays the guitar. He was born on September **Z**, 1949. The letter **Z** stands for the birth date. Use the clues to figure out **Z**.

CLUES:

1) 2 is not a factor of Z.
2) $Z \geq 15$
3) $Z \neq 21$
4) $3 \times Z \leq 25 + 25 + 25$
5) The sum of the digits of Z is 5 or less.

Bruce Springsteen was born on September ______ , 1949.

SOLVE IT: MENU MATTERS

How much is a bagel with cheese?

Betty's Bagels has 2 specials.
The signs are clues to the costs of the items.

SOLVE IT: ABC CODE CRACKERS

What is the value of B?

The numbers at the tops of the columns are the column sums.
The same letters have the same values. Crack the code.

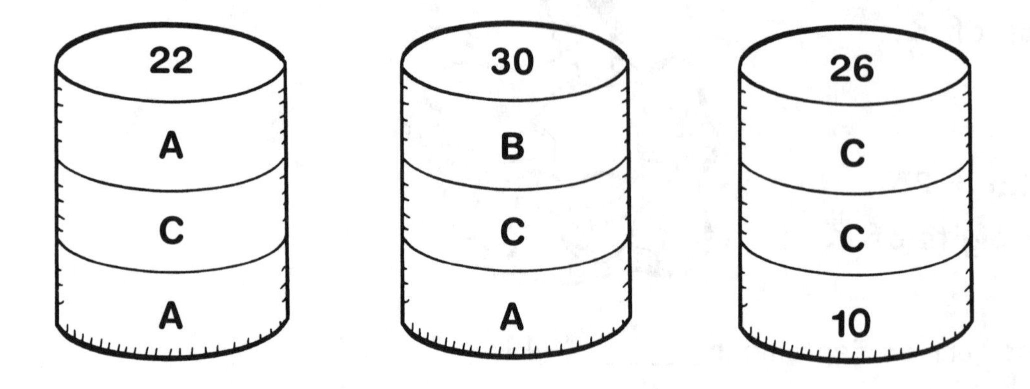

SOLVE IT: BALANCE THE BLOCKS

How many cylinders will balance 15 spheres?

All objects of the same shape are equal in weight.

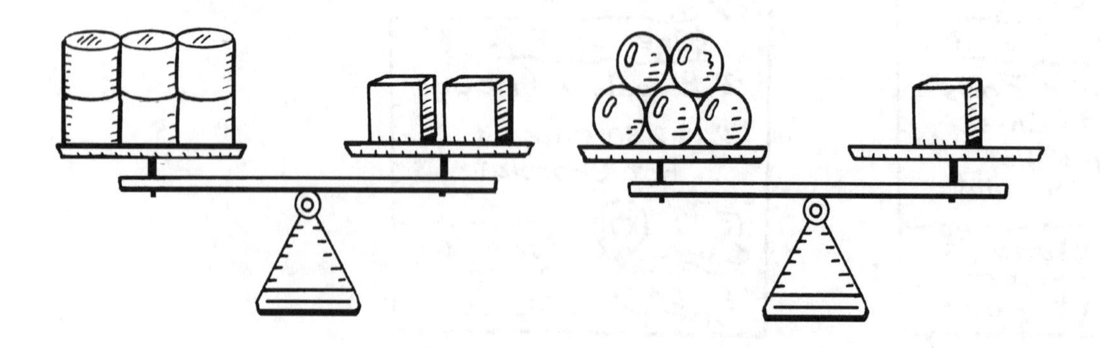

ANSWER KEY

Grid Patterns (pages 11–19)
Solve the Problem
1. 27, 28, 29
32, 33, 34
37, 38, 39
2. Numbers in rows are consecutive numbers. Numbers in columns differ by 5. The top row is 27, 28, 29. In the middle row, numbers are 5 more than the numbers above them, or 32, 33, 34. The bottom row follows the same pattern: 37, 38, 39. **3.** 39 **4.** $a + 12$

Make the Case
Who is on the ball? Buddy

Problem 1
1. 19, 20, 21
22, 23, 24
25, 26, 27
2. Numbers in the rows are consecutive numbers. Numbers in columns differ by 3. The bottom row is 25, 26, 27. In the middle row, numbers are 3 less than the numbers below them, or 22, 23, 24. The top row follows the same pattern: 19, 20, 21. **3.** 19 **4.** $b - 8$

Problem 2
1. 26, 27, 28
36, 37, 38
46, 47, 48
2. Numbers in rows are consecutive numbers. Numbers in columns differ by 10. The bottom row is 46, 47, 48. In the middle row, the numbers are 10 less than the numbers below them, or 36, 37, 38. The top row follows the same pattern: 26, 27, 28. **3.** 26 **4.** $c - 22$

Problem 3
1. 26, 27, 28
32, 33, 34
38, 39, 40
2. Numbers in rows are consecutive numbers. Numbers in columns differ by 6. The top row is 26, 27, 28. In the middle row, the numbers are 6 more than the numbers above them, or 32, 33, and 34. The bottom row follows the same pattern: 38, 39, 40. **3.** 40
4. $d + 14$

Problem 4
1. 34, 35, 36
42, 43, 44
50, 51, 52
2. Numbers in rows are consecutive numbers. Numbers in columns differ by 8. The top row is 34, 35, 36. In the middle row, the numbers are 8 more than the numbers above them or 42, 43, 44. The bottom row follows the same pattern: 50, 51, 52. **3.** 52 **4.** $e + 18$

Problem 5
1. 46, 47, 48
51, 52, 53
56, 57, 58
2. 46 **3.** $f - 12$ **4.** 35

Problem 6
1. 46, 47, 48
50, 51, 52
54, 55, 56
2. 56 **3.** $g + 5$ **4.** 64

Problem 7
1. 38, 39, 40
44, 45, 46
50, 51, 52
2. 38 **3.** $h - 7$ **4.** 55

Solve It: Grid Patterns
1. Look: In the array, numbers in columns differ from the numbers below them by 9. The problem is to figure out the greatest number in a 3-by-3 square in the array that has 56 as its middle number.
2. Plan and Do: Draw a 3-by-3 square. Record 56 in the middle square. Record the other two numbers in the middle row. The middle row is 55, 56, 57. Add 9 to each of these numbers to get the last row: 64, 65, 66.
3. Answer and Check: The greatest number is 66. To check, each number in a row should be 9 more than the number above it.

Dollar Dilemma (pages 22–30)
Solve the Problem
1. \$2 **2.** $\frac{1}{5}$ x 10, or \$2 **3.** \$16 **4.** \$9

Make the Case
Who is on the ball? Bobby

Problem 1
1. \$8 **2.** 6 + 2, or \$8 **3.** \$7 **4.** \$5

Problem 2
1. \$12 **2.** 8 + ($\frac{1}{4}$ x 16), or \$12 **3.** \$6
4. \$19

Problem 3
1. \$36 **2.** 32 + ($\frac{1}{10}$ x 40), or \$36
3. \$24 **4.** \$39

Problem 4
1. \$30 **2.** 26 + ($\frac{1}{6}$ x 24), or \$30
3. \$32 **4.** \$26

Problem 5
1. \$100 **2.** 60 + ($\frac{1}{3}$ x 120), or \$100
3. \$110 **4.** \$140

Problem 6
1. \$19 **2.** \$26 **3.** \$22 **4.** \$28

Problem 7
1. \$10 **2.** \$15 **3.** \$17 **4.** \$16

Solve It: Dollar Dilemma
1. Look: There are five penguins with phones and five facts about the cost of each phone. The problem is to figure out the cost of each phone.
2. Plan and Do: Work backward.
- Frosty's phone costs \$42.
- Paul's phone costs 30 + ($\frac{1}{3}$ x 42), or \$44.
- Sara's phone costs 18 + ($\frac{1}{2}$ x 44), or \$40.
- Tiptoe's phone costs 40 + 6, or \$46.
- Tiny's phone costs 46 + 40 = 86; $\frac{1}{2}$ x 86, or \$43.

3. Answer and Check: Tiny's cell phone costs \$43. To check, use the costs of each penguin's phone and check the costs with the facts. They must make sense.

Birthday Boggle (pages 33–41)
Solve the Problem
1. 5, 6, 7, ..., and 11 **2.** August 5, 1930 **3.** From Clues 1 and 2, M is 5, 6, 7, ..., or 11. Clue 3 eliminates all numbers except for 5 and 10. Clue 4 eliminates 10. So, M = 5. **4.** Clue 1: 3 x 5 ≥ 15. Clue 2: 5 < 12. Clue 3: 5 is a factor of 5 because 1 x 5 = 5. Clue 4: 5 ≠ 10.

Make the Case
Who is on the ball? Bobby

(Note: Explanations of solution methods may vary in the next set of problems.)
Problem 1
1. 1, 2, 3, 4, and 5 **2.** January 1, 1752 **3.** From Clue 1, R is 1, 2, 3, 4, or 5. Clue 2 eliminates the even numbers leaving 1, 3, and 5. Clue 3 eliminates 5. Clue 4 eliminates 3. So, R = 1. **4.** Clue 1: 1 x 1, or 1 ≤ 25. Clue 2: 1 is an odd number. Clue 3: 5 is not a factor of 1 because there is no whole number multiplied by 5 that gives 1 as a product. Clue 4: 1 ≠ 27 ÷ 9, or 3.

Problem 2
1. 22, 23, 24, ..., and 29 **2.** April 27, 1791 **3.** From Clues 2 and 3, C is 22, 23, 24, ..., or 29. Clue 4 eliminates 22, 23, 25, 26, 28, and 29, leaving 24 and 27. Clue 1 eliminates 24. So, C = 27. **4.** Clue 1: 4 is not a factor of 27 because there is no whole number multiplied by 4 that gives 27 as a product. Clue 2: 27 ≤ 29. Clue 3: 27 > 21. Clue 4: 3 is a factor of 27 because 9 x 3 = 27.

Problem 3
1. 16, 17, 18, 19, and 20 **2.** January 17, 1706 **3.** From Clues 1 and 5, G is 16, 17, 18, 19, or 20. Clue 2 eliminates 18 and 19, leaving 16, 17, and 20. Clue 3 eliminates 20. Clue 4 eliminates 16. So, G = 17. **4.** Clue 1: 17 ≤ 2 x 10, or 20. Clue 2: 1 x 7 = 7, and 7 is less than 8. Clue 3: 17 ≠ 9 + 11, or 20. Clue 4: 1 + 7, or 8, is not equal to 7. Clue 5: 17 ≥ 4 x 4, or 16.

Problem 4
1. February 12, 1809 **2.** From Clues 1 and 4, S is 10, 11, 12, ..., or 24. Clue 2 eliminates all numbers except for 12, 16, 20, and 24. Clue 3 eliminates 16 and 20, leaving 12 and 24. Clue 5 eliminates 24. So, S = 12. **3.** Clue 1: 12 > 18 ÷ 2, or 9. Clue 2: 4 is a factor of 12 because 3 x 4 = 12. Clue 3: 3 is a factor of 12 because 4 x 3 = 12. Clue 4: 12 + 12 ≤ 6 x 8, or 24 ≤ 48. Clue 5: 2 – 1 = 1, and 1 is less than 2. **4.** 1863

Problem 5
1. July 24, 1897
2. From Clues 2 and 5, F is 20, 21, 22, ..., or 30. Clue 3 eliminates all odd numbers. Clue 4 eliminates 20, 22, 26, and 28, leaving 24 and 30. Clue 1 eliminates 30. So, F = 24. **3.** Clue 1: 4 – 2, or 2, is not 3. Clue 2: 24 x 24, or 576 ≥ 400. Clue 3: 2 is a factor of 24 because 12 x 2 = 24. Clue 4: 3 is a factor of 24 because 8 x 3 = 24. Clue 5: 2 x 24, or 48, ≤ 60. **4.** 1932

Problem 6
1. May 29, 1917 **2.** From Clues 1 and 4, B is 26, 27, 28, ..., or 36. Clue 3 eliminates all even numbers. Clue 5 eliminates 31, 33, and 35, leaving 27 and 29. Clue 2 eliminates 27. So, B = 29. **3.** Clue 1: 29 + 29, or 58 > 50. Clue 2: 29 ≠ 23 + 4, or 27. Clue 3: 2 is not a factor of 29 because there is no whole number multiplied by 2 that has a product of 29. Clue 4: 29 ≤ 36. Clue 5: 2 x 9 = 18 and 18 is an even number. **4.** 1960

Problem 7
1. February 6, 1895 **2.** From Clues 3 and 4, Q is 5, 6, 7, ..., or 17. Clue 1 eliminates all numbers except for 6, 9, 12, and 15. Clue 5 eliminates 9 and 15, leaving 6 and 12. Clue 2 eliminates 12. So, Q = 6. **3.** Clue 1: 3 is a factor of 6 because 2 x 3 = 6. Clue 2: 6 ≠ 8 + 4, or 12. Clue 3: 6 x 6 ≥ 5 x 5, or 36 ≥ 25. Clue 4: 6 < 3 x 6, or 18. Clue 5: 2 is a factor of 6 because 3 x 2 = 6. **4.** 1969

Solve It: Birthday Boggle
1. Look: Five clues are given about the birth date of Bruce Springsteen. The number (birth date) is represented by the letter Z. Clues 2 and 4 give information about the least and the greatest values of Z.
2. Plan and Do. Clues 2 and 4 establish the range for Z; Z can be any number 15 through 25. Clue 1 indicates that 2 is not a factor of Z, so all even numbers are eliminated leaving 15, 17, 19, 21, 23, and 25. Clue 5 eliminates 15, 17, 19, and 25, leaving 21 and 23. Clue 3 eliminates 21. So, Z = 23. It is the only number that fits all of the clues.
3. Answer and Check: Z = 23. Replace Z with 23 and check 23 with each clue. Clue 1: 2 is not a factor of 23 because there is no whole number multiplied by 2 that gives 23 as a product. Clue 2: 23 ≥ 15. Clue 3: 23 ≠ 21. Clue 4: 3 x 23 ≤ 25 + 25 + 25, or 69 ≤ 75. Clue 5: 2 + 3, or 5, is 5 or less

Menu Matters (pages 44–52)
Solve the Problem
1. Since 3 chicken sandwiches + \$4.00 = \$22.00, then 3 chicken sandwiches = \$22.00 – \$4.00, or \$18.00 and each costs \$18.00 ÷ 3, or \$6.00. **2.** Replace each chicken sandwich with its cost in Special #1. Then (2 x \$6.00) + 2 ears of corn = \$16.00, and the 2 ears of corn are \$16.00 – \$12.00, or \$4.00. Each ear of corn is \$4.00 ÷ 2, or \$2.00. **3.** Special #1: (2 x \$6.00) + (2 x \$2.00) = \$12.00 + \$4.00, or \$16.00. Special #2: (3 x \$6.00) + \$4.00 = \$18.00 + \$4.00, or \$22.00.

Make the Case
Who is on the ball? Buddy

Problem 1
1. In Special #1, the total cost of 3 banana smoothies is \$12.00. Each banana smoothie is \$12.00 ÷ 3, or \$4.00. **2.** In Special #1, a banana smoothie is \$12.00 ÷ 3, or \$4.00. In Special #2, replace the banana smoothie with its cost. Then the 2 orange smoothies are \$10.00 – \$4.00, or \$6.00, and each orange smoothie is \$6.00 ÷ 2, or \$3.00. **3.** Special #1: 3 x \$4.00 = \$12.00. Special #2: \$4.00 + (2 x \$3.00) = \$4.00 + \$6.00, or \$10.00.

Problem 2
1. In Special #2, the total cost of 2 peanut butter and jelly sandwiches without the cookies is \$11.00 – \$3.00, or \$8.00, and each sandwich is \$8.00 ÷ 2, or \$4.00. **2.** In Special #2, a peanut butter and jelly sandwich is \$4.00. In Special #1, replace the peanut butter and jelly sandwich with its cost. Then the dish of ice cream is \$7.00 – \$4.00, or \$3.00. **3.** Special #1: \$4.00 + \$3.00 = \$7.00. Special #2: \$3.00 + (2 x \$4.00) = \$3.00 + \$8.00, or \$11.00.

Problem 3
1. In Special #1, the total cost of 4 onion rolls is \$27.00 – \$3.00, or \$24.00, and each onion roll is \$24.00 ÷ 4, or \$6.00. **2.** In Special #1, an onion roll is \$6.00. In Special #2, replace each onion roll with its cost. Then the 2 roast beef sandwiches are \$30.00 – (2 x \$6.00), or \$18.00. Each roast beef sandwich is \$18.00 ÷ 2, or \$9.00. **3.** Special #1: (4 x \$6.00) + \$3.00 = \$24.00 + \$3.00, or \$27.00. Special #2: (2 x \$9.00) + (2 x \$6.00) = \$18.00 + \$12.00, or \$30.00.

Problem 4
1. In Special #2, the total cost of 2 hamburgers without the orange sodas is \$17.00 – (2 x \$1.50), or \$14.00 and

each is $14.00 ÷ 2, or $7.00. **2.** In Special #2, a hamburger is $7.00. In Special #1, replace the hamburger with its cost. Then the 3 club sandwiches are $31.00 – $7.00, or $24.00, and each one is $24.00 ÷ 3, or $8.00. **3.** Special #1: $7.00 + (3 x $8.00) = $7.00 + $24.00, or $31.00. Special #2: (2 x $7.00) + (2 x $1.50) = $14.00 + $3.00, or $17.00.

Problem 5

1. $6.00 **2.** In Special #1, a salad rollup is $24.00 ÷ 4, or $6.00. In Special #2, replace each rollup with $6.00. Then the total cost of 2 glasses of cider is $22.00 – (2 x $6.00) – $2.00 = $22.00 – $12.00 – $2.00, or $8.00, and each glass of cider is $8.00 ÷ 2, or $4.00. **3.** Special #1: 4 x $6.00 = $24.00. Special #2: (2 x $6.00) + (2 x $4.00) + $2.00 = $12.00 + $8.00 + $2.00, or $22.00.

Problem 6

1. $4.50 **2.** $2.50 **3.** In Special #1, the total cost of 3 hot dogs without the ice cream is $16.00 – $2.50, or $13.50, and each hotdog is $13.50 ÷3, or $4.50. In Special #2, replace each hot dog with its cost. Then the 3 baked potatoes are $18.50 – (2 x $4.50) – $2.00, or $7.50. Each baked potato is $7.50 ÷ 3, or $2.50.

Problem 7

1. $2.00 **2.** $1.50 **3.** In Special #1, the total cost of 3 eggs without the milk is $8.00 – $2.00, or $6.00 and each egg is $6.00 ÷ 3, or $2.00. In Special #2, the total cost of 2 pancakes is $12.00 – (2 x $2.00) – $3.00 – $2.00, or $3.00 and each pancake is $3.00 ÷ 2, or $1.50.

Solve It: Menu Matters

1. Look: There are two specials. Special #1 is 1 bagel with eggs, 1 bagel with cheese, and a $2.50 cup of hot chocolate for a total cost of $9.00. Special #2 is two bagels with eggs and two $2.50 cups of hot chocolate for a total of $12.00. The problem is to figure out the cost of a bagel with cheese.

2. Plan and Do: In Special #2, the total cost of 2 bagels with eggs is $12.00 – (2 x $2.50), or $7.00, and each bagel with eggs is $7.00 ÷ 2, or $3.50. In Special #1, replace the bagel with eggs with its cost. Then the bagel with cheese is $9.00 – $3.50 – $2.50, or $3.00.

3. Answer and Check: The bagel with cheese is $3.00. To check, replace each item in the specials with its cost. Add the costs and compare with the total costs given in the specials. Special #1: $3.50 + $3.00 + $2.50 = $9.00. Special #2: (2 x $3.50) + (2 x $2.50) = $12.00.

ABC Code Crackers (pages 55–63)

Solve the Problem

1. Starting with the second column, she can figure out that A + B = 19 – 6, or 13. Replacing A and B in the first column with 13, she can get the value of C.**2.** 7 **3.** 9 **4.** In the second column, 6 + B + A = 19. So, B + A = 19 – 6, or 13. In the first column, replace A and B with 13. Then C is 20 – 13, or 7. In the third column, replace each C with 7. Then A = 23 – 7 – 7, or 9. In the second column, replace A with 9. Then B = 19 – 6 – 9, or 4.

Make the Case

Who is on the ball? Buddy

Problem 1

1. There's only one letter, so C + C = 19 – 3, or 16, and C = 16 ÷ 2, or 8. By replacing C with 8 in the first column, she can figure out the value of A. **2.** 10 **3.** In the third column, C = 8. In the first column, replace C with 8. Then A + A = 28 – 8, or 20, and A is 20 ÷ 2, or 10. In the second column, replace A with 10 and C with 8. Then B = 24 – 10 – 8, or 6. **4.** 48

Problem 2

1. She can figure out that A + A = 35 – 11, or 24, and A is 24 ÷ 2, or 12. By replacing A with 12 in the third column, she can figure out the value of B. **2.** 8 **3.** In the first column, A + A = 35 – 11, or 24. A = 24 ÷ 2, or 12. In the third column, replace A with 12. Then B + B = 28 – 12, or 16 and B = 16 ÷ 2, or 8. In the second column, replace A with 12 and B with 8. Then C = 29 – 8 – 12, or 9. **4.** 41

Problem 3

1. She can figure out that B + C = 23 – 9, or 14. **2.** 2 **3.** From the third column B + C = 23 – 9, or 14. In the first column, replace B and C with 14. Then A = 16 – 14, or 2. In the second column, replace A with 2. Then C + C = 8 – 2, or 6, and C = 6 ÷ 2, or 3. **4.** 31

Problem 4

1. 18 **2.** 9 **3.** In the second column, A + C = 22 – 4, or 18. In the third column, replace A and C with 18. Then B = 27 – 18, or 9. In the first column, replace B with 9. Then A = 26 – 9 – 9, or 8. **4.** 15

Problem 5

1. 30 **2.** 7 **3.** 20 **4.** In the first column, A + C = 38 – 8, or 30. In the third column, replace A and C with 30. Then B = 37 – 30, or 7. In the second column, replace B with 7. Then A + A = 47 – 7, or 40 and A = 40 ÷ 2, or 20. In the first column, replace A with 20. Then C = 38 – 20 – 8, or 10.

Problem 6

1. 20 **2.** 10 **3.** 15 **4.** In the second column, B + 6 + C = 26, so B + C = 26 – 6, or 20. In the first column, replace B and C with 20. Then A = 30 – 20, or 10. In the third column, replace A with 10. Then B + B = 40 – 10, or 30, and B = 30 ÷ 2, or 15. In the second column, replace B with 15. Then C = 26 – 15 – 6, or 5.

Problem 7

1. 6 **2.** 19 **3.** 15 **4.** In the first column, A = 6. In the second column, replace A with 6. Then B + C = 19. In the third column, replace B and C with 19. Then the other B is 34 – 19, or 15. In the third column, replace each B with 15. Then C = 34 – 15 – 15, or 4.

Solve It: ABC Code Crackers

1. Look: There are three columns of letters and numbers. The numbers on the tops of the columns are the sums of the numbers and values of the letters in the columns. The first column sum is 22; the second column sum is 30; and the third column sum is 26. There are three different letters. The third column contains the number 10.

2. Plan and Do: Subtract 10 from the sum in the third column. Then C + C = 16 and C = 16 ÷ 2, or 8. Replace C with 8 in the first column. Then A + A = 22 – 8, or 14, and A = 14 ÷ 2, or 7. Replace C and A with their values in the second column. Then B = 30 – 8 – 7, or 15.

3. Answer and Check: B= 15; To check, replace each letter in the columns with its value and add. Check the sums with the numbers on the tops of the columns. First column: 7 + 8 + 7 = 22. Second column: 15 + 8 + 7 = 30. Third column: 8 + 8 + 10 = 26.

Balance the Blocks (pages 66–74)

Solve the Problem

1. Ima started with the first pan balance because she could figure out that 2 spheres balance 1 cylinder. She can then substitute 2 spheres for the cylinder in the second pan balance. **2.** 2 **3.** 6 **4.** Possible answer: In the first pan balance, since 4 spheres balance 2 cylinders, 2 spheres (4 ÷ 2) will balance 1 cylinder (2 ÷ 2). In the second pan balance, 1 cylinder balances 3 cubes. Substitute 2 spheres for the 1 cylinder. That means that 2 spheres will balance 3 cubes, and 6 spheres (3 x 2) will balance 9 cubes (3 x 3). **5.** 3 pounds

Make the Case

Who is on the ball? Sally

Problem 1

1. Ima started with the first pan balance because she could figure out that 2 spheres will balance 1 cube. Then she could substitute 2 spheres for each cube in the second pan balance. **2.** 8 **3.** 16 **4.** Possible answer: In the first pan balance, 4 spheres balance 2 cubes, so 2 spheres (4 ÷ 2) will balance 1 cube (2 ÷ 2). In the second pan balance, substitute 2 spheres for each cube. Since 8 spheres will balance 2 cylinders, then 16 spheres (2 x 8) will balance 4 cylinders (2 x 2). **5.** 8 pounds

Problem 2

1. Ima started with the second pan balance because she could figure out that 2 spheres will balance 1 cylinder. Then she could substitute 2 spheres for each cylinder in the first pan balance. **2.** 12 **3.** 24 **4.** Possible answer: In the second pan balance, 6 spheres balance 3 cylinders, so 2 spheres (6 ÷ 3) will balance 1 cylinder (3 ÷ 3). In the first pan balance, substitute 2 spheres for each cylinder. Since 12 spheres will balance 2 cubes, then 24 spheres (2 x 12) will balance 4 cubes (2 x 2). **5.** 6 pounds

Problem 3

1. Ima started with the second pan balance because could figure out that 2 cubes will balance 1 cylinder. Then she could substitute 2 cubes for each cylinder in the first pan balance. **2.** 6 **3.** 4 **4.** Possible answer: In the second pan balance, 2 cylinders balance 4 cubes, so 1 cylinder (2 ÷ 2) will balance 2 cubes (4 ÷ 2). In the first pan balance, substitute 2 cubes for each cylinder. Since 9 spheres balance 6 cubes, then 3 spheres (9 ÷ 3) will balance 2 cubes (6 ÷ 3), and 6 spheres (2 x 3) will balance 4 cubes (2 x 2). **5.** 4 pounds

Problem 4

1. 1 **2.** 8 **3.** 12 **4.** Possible answer: In the first pan balance, 4 cylinders balance 1 cube. In the second pan balance, substitute 4 cylinders for each cube. Since 8 cylinders will balance 6 cubes, then 4 cylinders (8 ÷ 2) will balance 3 spheres (6 ÷ 2). Then 12 cylinders (3 x 4) will balance 9 spheres (3 x 3). **5.** 2 pounds

Problem 5

1. 2 **2.** 10 **3.** 15 **4.** Possible answer: In the second pan balance, 4 cylinders balance 2 spheres, so 2 cylinders (4 ÷ 2) will balance 1 sphere (2 ÷ 2). In the first pan balance, substitute 2 cylinders for the sphere. Since 5 cubes balance 2 cylinders, then 15 cubes (3 x 5) will balance 6 cylinders (3 x 2). **5.** 2 pounds

Problem 6

1. 1 **2.** 6 **3.** 9 **4.** Possible answer: In the second pan balance, 3 cylinders balance 1 sphere. In the first pan balance, substitute 3 cylinders for each sphere. Since 4 cubes will balance 6 cylinders (3 x 2), then 2 cubes (4 ÷ 2) will balance 3 cylinders (6 ÷ 2), and 6 cubes (3 x 2) will balance 9 cylinders (3 x 3). **5.** 4 pounds

Problem 7

1. 4 **2.** 16 **3.** 12 **4.** Possible answer: In the first pan balance, 3 cylinders balance 6 spheres, so 1 cylinder (3 ÷ 3) will balance 2 spheres (6 ÷ 3). In the second pan balance, substitute 2 spheres for each cylinder. Since 8 cubes balance 4 spheres (2 x 2), then 4 cubes (8 ÷ 2) will balance 2 spheres (4 ÷ 2), and 12 cubes (3 x 4) will balance 6 spheres (3 x 2). **5.** 3 pounds

Solve It: Balance the Blocks

1. Look: In the first pan balance, 6 cylinders balance 2 cubes. In the second pan balance, 5 spheres balance 1 cube. The problem is to figure out how many cylinders will balance 15 spheres.

2. Plan and Do: In the first pan balance, 6 cylinders balance 2 cubes, so 3 cylinders (6 ÷ 2) will balance 1 cube (2 ÷ 2). In the second pan balance, substitute 3 cylinders for the cube. Then 5 spheres balance 3 cylinders, and 15 spheres (3 x 5) will balance with 9 cylinders (3 x 3).

3. Answer and Check: 9 cylinders will balance 15 spheres. To check, suppose that 1 cube weighs 15 pounds. Then 1 sphere will be 3 pounds (15 ÷ 5) because 5 x 3 = 3 x 5. Then 2 cubes would weigh 2 x 15, or 30 pounds. Then 6 cylinders would weigh 30 pounds, and each cylinder would weigh 5 pounds (30 ÷ 6). Then 15 spheres would weigh 45 pounds (15 x 3) and 9 cylinders would weigh 45 pounds (9 x 5). Since 15 spheres weigh the same as 9 cylinders, the answer is correct.